GALLJAMESBURKERICHARDFEYNMAN
RROTIMAXJAMMERLEWISMU D
UJANNALEVINANTONI BIM
ISTOTLEHENRIMATISSEPIETMONDRIAN
NEDWARDHOPPERHAYDENWHITE
EDSTIEGLITZGARYZUKAVPEARLBUCK
ERNICUSERWINSCHRÖDINGERHGWELLS
HKATSUSHIKAHOKUSAIDYLANTHOMAS
LDECERVANTESHELENFRANKENTHALER
SSOSIRISAACNEWTONPAULGAUGUIN
BECKERJOANMIRÓSTEVENWEINBERG
UDWIGVANBEETHOVENSIRJAMESJEANS
SATTMARCELPROUSTGEORGESSEURAT
TOGIACOMETTINICCOLÒMACHIAVELLI
STEAURACHELCARSONALBERTCAMUS
QUEZWALTERBENJAMINRENÉMAGRITTE
RTHESJOHNRUSKINEMILEBERNARD
AUDEBROWNROBERTMOTHERWELL
DESCARTESDOUGLASHOFSTADTER
INGWAYHENRIDETOULOUSE-LAUTREC
MAURICEDENISSTEPHENHAWKING
HUTEJOHANNWOLFGANGVONGOETHE
SGLEICKJOHANNESKEPLERGINOSEGRÈ
NDRSOLZHENITSYNALANLIGHTMAN
LETTOSTUDSTERKELHOWARDZINN
GIOTTOALEXISDETOCQUEVILLEHOMER
JOHNALTONPAULDIRACARCHIMEDES
EBRORDSNELLIUSRICHARDRHODES

DR. HELEN CALDICOTT
FOUNDER AND PRESIDENT OF
THE NUCLEAR POLICY RESEARCH INSTITUTE

"*Categories—On the Beauty of Physics* lays open the elegance of physics to the appreciation of the lay-scientist. By trying to understand the universe in which we live, we are all basically plain-clothed scientists learning the tools that are necessary to save the planet. That is why I think this book is important as an educational and recreational piece of work. Pick it up. Explore the different manifestations of important physics concepts in art and literature. *Categories—On the Beauty of Physics* facilitates the appreciation of the beauty of these concepts in their scientific form. This wonderful book will provoke thought in lovers of science and art alike, and, with knowledge comes the inspiration to preserve the beauty of life on Earth."

JOHN RENNIE
EDITOR-IN-CHIEF, *SCIENTIFIC AMERICAN*

"Ideas can have a life of their own, and as *Categories—On the Beauty of Physics* elegantly proves, the concepts native to science feed the imagination as much as they draw from it. Thoughts of particles, for example, wing their way from subatomic blurs to Seurat's pointillism, to Proust's memorable madeleines and back to the mysterious dark matter defining most of the universe. The book is a treat for scientists and poets alike."

JOHN KATZMAN
FOUNDER OF THE PRINCETON REVIEW

"Everyone talks about the beauty and elegance of physics, but this is the first beautiful and elegant physics book."

JOHN GRIBBIN
AUTHOR OF *THE NEW SCIENTISTS* & *IN SEARCH OF SCHRÖDINGER'S CAT*

"People who would be ashamed to admit that they had never heard of William Shakespeare, or the origin of the quotation 'To be, or not to be' will often proudly proclaim that they know nothing at all about science. Yet science is as significant and important a part of the human achievement as literature, and there are certain scientific ideas that ought to be part of the currency of intelligent discussion. Some of those ideas are presented here, in a form which will do much to break down the idea that the concepts of science are always difficult to grasp. We cannot all write like Shakespeare, but we can all appreciate the beauty of his sonnets; in the same way, we cannot all be Nobel-prize winning physicists, but we can all get a glimpse of what it is those physicists have achieved. And this book is an excellent place to begin that process of understanding."

ALAN LIGHTMAN
AUTHOR OF *EINSTEIN'S DREAMS*

"A beautiful synthesis of science and art, pleasing to the mind and to the eye."

STEVEN PINKER
AUTHOR OF *HOW THE MIND WORKS* & *THE LANGUAGE INSTINCT*

"A gorgeous book—proof that beauty can be found in equal measure in words, images, and ideas."

JACQUES D'AMBOISE
FOUNDER OF THE NATIONAL DANCE INSTITUTE

"A feast of discovery, a symphony of learning! Open to any page of *Categories—On the Beauty of Physics* and prepare to be filled with a sense of wonder about the range and breadth of human accomplishments, including those of Hilary Hamann and her colleagues. This fusion of science and art confirms that creative approaches to learning are thrilling and inspiring, and that art remains the best vehicle through which to transform minds and lives—both young and old! This is a book to treasure."

JAMES BURKE
AUTHOR AND PRODUCER OF *CONNECTIONS*

"An extraordinary, beautiful, and stimulating book. The physics texts are jewels of descriptive clarity and, in the weave of science with the arts, there are moments of true revelation. In a world of growing interdependence, this book is a must-read (if only to discover that 'a tomato is every color but red'!)"

ALAN CHODOS
AMERICAN PHYSICAL SOCIETY & WORLD YEAR OF PHYSICS 2005

"*Categories—On the Beauty of Physics* is a unique work that suggests deep connections among physics, literature, and art. The classical physics text is entertaining and pedagogically sound, though the approach of treating topics alphabetically makes it more like an encyclopedia than a textbook. The carefully chosen literary quotations and visual artworks illuminate and extend the physics in ways that entice the mind to wander outside the boundaries of recognized academic disciplines. A valuable and fascinating book."

ROBERT EHRLICH
AUTHOR OF *WHY TOAST LANDS JELLY-SIDE DOWN*

"This book gives some very nice, easy to understand explanations of a wide variety of ideas in modern physics."

HENRY PETROSKI
AUTHOR OF *TO ENGINEER IS HUMAN* & *PUSHING THE LIMITS*

"*Categories—On the Beauty of Physics* is a remarkable example of what wondrous things can come of a fruitful collaboration among scientists, writers, and artists. This book is an outstanding testament to the inherent interdependence of all human thought and creativity, and its whole is truly greater than the sum of its parts."

MARTIN A. UMAN
AUTHOR OF *ALL ABOUT LIGHTNING*

"This is a remarkable and unique book."

GINO SEGRÈ
AUTHOR OF *A MATTER OF DEGREES*

"*Categories—On the Beauty of Physics* is a strikingly handsome and interesting combination of science, art, and literature, educating and enchanting the reader at the same time as it points to myriad new paths for exploration. A different and interesting new kind of venture."

V.

CATEGORIES
ON THE BEAUTY OF PHYSICS

ESSENTIAL PHYSICS CONCEPTS AND THEIR COMPANIONS IN ART AND LITERATURE

CATEGORIES
BOOK ONE

AN H. T. HAMANN BOOK

SCIENCE WRITER & EDITOR—EMILIANO SEFUSATTI, PH.D.
COLLAGE ARTIST—JOHN MORSE
FINE ART & LITERATURE REVIEWER—HILARY THAYER HAMANN, M.A.

THE CATEGORIES SERIES
Cultivating the Connected Mind™

Vernacular Press

CATEGORIES—ON THE BEAUTY OF PHYSICS

AN H. T. HAMANN BOOK
FIRST EDITION

published by
VERNACULAR PRESS

V.

PUBLISHED IN THE UNITED STATES OF AMERICA BY VERNACULAR PRESS, NEW YORK.

LIBRARY OF CONGRESS CATALOG NUMBER: 2005930500
ISBN: 0-9740266-3-8

10 9 8 7 6 5 4 3 2 1

ACKNOWLEDGMENTS FOR PREVIOUSLY PUBLISHED MATERIALS AND FINE ART CAN BE FOUND AT THE CLOSE OF THE BOOK.

PRINTED IN THE UNITED STATES OF AMERICA
UNDER THE TECHNICAL SUPERVISION OF JAMES BENARD
OF MATCH FINE PRINT, NEW YORK, NEW YORK,
WITH RETOUCHING BY CHRISTIAN MCGRATH.

THE CATEGORIES SERIES
Cultivating the Connected Mind™
IS A REGISTERED TRADEMARK.

VERNACULAR PRESS
560 BROADWAY, SUITE 509
NEW YORK, NY 10012
WWW.VERNACULARPRESS.COM

CONTRIBUTORS

SCIENCE WRITER & EDITOR
EMILIANO SEFUSATTI, PH.D.

COLLAGE ARTIST
JOHN MORSE

FINE ART & LITERATURE REVIEWER
HILARY THAYER HAMANN, M.A.

CONTRIBUTING SCIENCE WRITERS
STEPHEN MARKACS,
ALVARO NÚÑEZ NIKITIN, PH.D., & GABRIJELA ZAHARIJAS, PH.D.

PROJECT MANAGER & ART DIRECTOR
CHRISTINE VECOLI

MANAGING EDITOR
MATT COLABRARO

EDITOR
HILLERY HUGG

PROJECT DEVELOPMENT
MARY PETERS

EDUCATIONAL ADVISORS
JOVAN FRANCHETTI, PATRICIA THAYER HOPE, M.S., & DORIAN MCCORMICK, M.ED.

EDITORIAL ASSISTANTS
LIZ AFTON, EMILY ALDRIDGE, VEE THAYER BENARD, LYDIA BELL,
CECILE BERBERAT, MELISSA BROWN, ALI BUJNOWSKI, MEGHAN-MICHELE GERMAN,
ANDREW HIRSHMAN, JULIA HOLMES, PENELOPE HOPE, MATT LYNCH,
CAROL MASKUS, THOMAS PASQUINI, DONNA SHERWOOD

THIS BOOK HAS DEPENDED ON THE ENCOURAGEMENT, ENTHUSIASM, AND ADVICE OF COUNTLESS CONCERNED AND DEDICATED INDIVIDUALS. SPECIAL THANKS TO THE PROFESSIONAL PHYSICISTS TO WHOM WE SUBMITTED THE MATERIAL FOR THEIR GENEROUS ATTENTION AND SOUND COUNSEL, AND TO THE EAST HAMPTON HIGH SCHOOL SCIENCE DEPARTMENT, EAST HAMPTON, NEW YORK, IN PARTICULAR TO PHYSICS TEACHER DAVID WICKS AND TO MASTER TEACHER AND SCIENCE CHAIRPERSON PATRICIA THAYER HOPE FOR KIND CONSULTATION.

THE CATEGORIES SERIES

PUBLISHER
JAMES BENARD

CREATIVE & EDITORIAL DIRECTOR
HILARY THAYER HAMANN, M.A.

PROJECT MANAGER & ART DIRECTOR
CHRISTINE VECOLI

MANAGING EDITOR
MATT COLABRARO

DEDICATION

FOR JACQUES

Years ago, life afforded me the opportunity to work with the most extraordinary individual under the most extraordinary circumstances. It was my privilege to serve as the assistant to Jacques d'Amboise, the founder and then artistic director of the National Dance Institute, a non-profit organization dedicated to the belief that the arts have the unique capacity to engage children and motivate them toward excellence. NDI, which works with thousands of children each year in schools throughout New York City and in communities nationwide, achieves so much that it almost defies description. In the most formal regard, the organization is dedicated to improving the lives of children through the magic of dance and the thrill of performance, and through encounters with creative professionalism that are indelible. In a less formal but arguably more important regard, the endeavor is about the willingness and the determination to "reach." In the way that a visually-impaired person must feel his or her way through a shadowy world, NDI reaches out of necessity—it reaches to touch. From Jacques and from my former colleagues, many of whom still belong to the company, I learned "to reach to touch." If the Categories series succeeds in any small measure to captivate readers and encourage them to explore the sciences and the humanities, it is to my own indelible encounter with the arts that I will owe that success.

It is with great pleasure that I dedicate this book
to my friend Jacques d'Amboise, Artist and Master Teacher,
and Expert in the Joy of Giving
for teaching me that originality is everywhere,
that to get the best, we must expect the best,
and that, above all things, we must remain impressionable.

—Hilary Thayer Hamann

CONTENTS

THE CATEGORIES SERIES
Cultivating the Connected Mind

Categories—On the Beauty of Physics is the first in a series of multidisciplinary, interdisciplinary educational books that will use imagery from emerging and well-known artists, and information from a variety of disciplines—*categories*—to facilitate the reader's encounter with challenging material. *Categories—On the Beauty of Physics* is not a physics textbook; it is a book *about* physics that uses literature and art to stimulate the wonder and interest of the reader. It is intended to promote scientific literacy, foster an appreciation of the humanities, and encourage readers to make informed and imaginative connections between the sciences and the arts. The book can be used as a cooperative learning tool through which people (especially educators and students) can engage in academic and value-oriented discussions. The *Categories* series was conceived and realized by recognizing that content compiled in an original and visually-compelling manner has the capacity to:

- inspire early learners to become lifelong learners
- inspire late learners to take a renewed interest in material previously believed to be too difficult
- improve the academic performance of students by encouraging enthusiastic adult involvement in student learning
- improve the quality of family life by giving families clear, well-researched tools with which to discuss ideas and connect meaningfully
- improve the quality of academic experience for American students from various backgrounds by providing young people with "gateway" materials to the arts, sciences, and humanities
- foster an atmosphere of intelligent play in the classroom, at home, or simply among friends

CATEGORIES ON THE BEAUTY OF PHYSICS
How to Use this Book

1. FEATURED CONCEPT

Each of the thirty-nine alphabetized chapters is dedicated to a concept taken from classical or modern physics. Individually, these chapters will offer a unique and thought-provoking sketch of the featured term. If read in succession, the chapters will provide a basic understanding of physics, past and present.

2. FEATURED LITERARY QUOTE

How do authors describe *motion* as sound, *angles* as historical interpretation, or *velocity* as a life of chaotic events and wasted potential? Here, the reader becomes acquainted with significant literary and academic works and is invited to imagine the expository possibilities of the scientific terms.

3. DICTIONARY DEFINITION

A selection from a dictionary definition demonstrates the important difference between the meaning of the selected word in common language and in the language of physics.

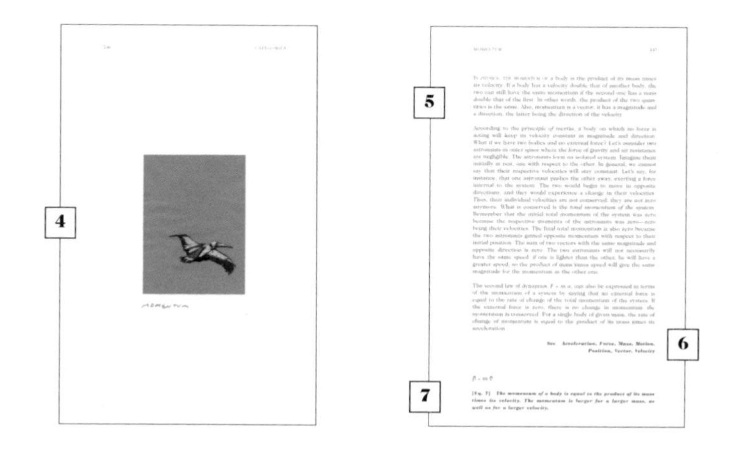

4. COLLAGE ART

Visually stimulating images invite the reader to grasp the essence of the physics concepts. These are not scientific diagrams; rather, they are intended to bring the spirit of scientific themes memorably to life.

5. PHYSICS TEXT

In the space of a single page, a physicist offers an innovative conceptual overview of the featured term, intended to introduce nonscientists to the beauty of physics, while providing more experienced learners with creative reinforcement of existing ideas.

6. SEE

This list refers the reader to related terms featured in the book and emphasizes relationships between key concepts.

7. EQUATIONS

Frequently, equations are provided in order to show the specialized way in which physicists illustrate complex ideas.

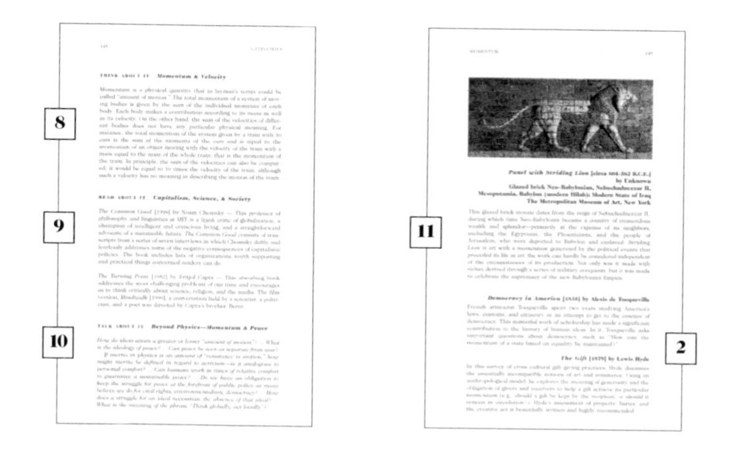

8. THINK ABOUT IT

This section supplements the physicist's description provided on the preceding page. Pertinent themes related to the term are emphasized for clarity.

9. READ ABOUT IT

Categories offers approximately one hundred reviews of classic and contemporary books, films, and articles in order to direct interested readers to related works of merit. All recommended materials can be found in libraries, in bookstores, or on-line. None of these reviews were solicited by or paid for by the publishers or authors.

10. TALK ABOUT IT

This constellation of concepts will inspire creative brainstorming. Readers are invited to use the featured term as a starting place for open-ended thinking and conversation. Note: there are no *correct* answers to these questions!

11. FEATURED FINE ART

The fine art selections will contribute to the depth of the reader's impression of physics. Though the art can be seen on-line, readers are encouraged to visit museums and galleries whenever possible.

INTRODUCTION

Simple Beginnings

The genesis of this project is simple. In the autumn of 2003, my friend John Morse displayed several of his collages at the Vernacular gallery. I am an admirer of John's work, so I loved all of the pieces shown, but one collage in particular captured my imagination: a torn paper rendering of the Roman Coliseum. I found the work to be elementary and sublime—it reminded me, most definitively, of entropy. When the show came down, John kindly left the collage, and I hung it over my desk, where, with a quiet insistence, it impressed itself upon me. The image is the representation of an icon—an icon of an icon, so to speak—and, as such, is densely packed with meaning. The Coliseum itself is a place of extremes. It is an extant historical marker, a confirmation of the enduring presence of the past. It refers to grandiose beginnings (gladiatorial combat, human and animal sacrifice, mock naval battles) and pitiful ends (the fall of Rome). There is difficulty in trying to separate the original political design from the extrinsic effects of decay; the meaning of the place has been co-opted by time's inexorable effects. The Coliseum has been transformed, and it continues to be transformed, moving from one state to another, from something more or less synthetic to something more or less particulate. The edifice's disrepair is itself perfectly iconic: the former sign of imperial dominance has become a symbol of the far greater might of material impermanence.

Over the course of the next several months, I began to think about entropy more and more frequently, to the exclusion of other, previously pressing things. Whenever I would take a break from my writing, I would look at the collage. The more I looked, the more I thought. Yet, the more I thought, the less I felt I knew. My uncertainty pertained to entropy and *directionality* and entropy and *intentionality*. I wondered about entropy and a linear conception of time, and about entropy and manufactured versus organic forms (a steam engine versus the human genome). I wondered about entropy and the differences between form and function (the *purpose* of a steam engine versus the *purpose* of a genome), and between functional and detrimental disorder (the shuffling of a deck of cards for maximum randomness/fairness versus the unstoppable deterioration of Leonardo da Vinci's *The Last Supper*). I wondered what to make of the negative connotations of words like *decay*, *disrepair*, and *disorder*, when, in fact, the processes associated with these words are not necessarily negative. And isn't the functional disorder ascribed to a shuffled deck of cards relevant only to the *humans* using the cards, and the detrimental disorder applied to a diminishing masterpiece relevant only to the *humans* who appreciate the art? No matter how I approached the topic, I could not escape the subjectivity of human experience, and I began to wonder if there might be a "correct" way to think about entropy.

I decided to do some reading, and I began with familiar pieces: "Entropy and Art," an essay I'd read in graduate school by theorist Rudolf Arnheim that addresses the contradiction between the tendency in humans and in nature towards order (as described by evolution) and the tendency in the universe towards disorder (as described by the second law of thermodynamics); and Percy Bysshe Shelley's "Ozymandias," which I hadn't read since high school but that I remembered vividly because of the effect that its portrait—and confession—of the pointlessness of human intention had had on me. Ladislao Reti's *The Unknown Leonardo* provided a wealth of details on the text and context of Leonardo's *The Last Supper*, which I was fortunate enough to see with my daughter and my younger sister (then ages 2 and 9, respectively) on a trip to Milan. From my collection of science books, I learned that physics gives entropy a precise definition—*the ratio of heat to temperature*—and yet, although entropy has a formula and can be measured, I found that the principle tends to capture the imagination in ways more thematic than literal. Within the modest confines of my research, entropy was discussed in relation to chaos, probability, economics, Darwinism, the heat death of the universe, the aesthetic representation of space-time, black holes, magic cubes, information, the messiness of a child's playroom, and even test patterns on television.[1]

An Invitation to Learn

Clearly, my collage of the Coliseum is not some consummate symbol of entropy as physics describes it. Like many works of art, the image simply constitutes an invitation to dream and, perhaps, to act. By allowing myself to be moved by it, I embarked on an adventure in learning on which I was introduced to great thoughts of great minds. The art did not take the place of serious reading in science; rather, it provided the *enhancement of* and the *impetus toward* serious reading in science. Since it was not my goal to become an expert, I enjoyed the luxury of reading broadly. I became absorbed *enough* in discussions of controlled versus random arrangements, natural versus unnatural tendencies, reversible versus irreversible processes, and open versus closed systems, to test and modify my own thinking, to feel authorized to seek new information on physics in particular and on the sciences in general, and to initiate conversations with family and friends, many of whom are scientists. Most important, and somewhat unexpectedly, I felt inspired and better able to encourage my children in their scientific interests and pursuits.

A Tool for Teaching and Learning

I began to wonder: would it be possible to curate similar explorations in science using art and literature? Could I re-create for beginners the sensation of impressionability and self-determination I'd experienced? Could pictures and words be presented in a way that is complementary, but not expected; connected, yet only indirectly; familiar, but completely new? Could I produce a book that would be fresh enough to arouse the wonder of the young, and sophisticated enough to invigorate—or *re*invigorate—the enthusiasm of the more mature? Could that book be used as a tool for cooperative engagement; that is, could people—especially adults and children—read it together? Could it be seen (at least in small measure) as fortification against the declining interest in the sciences among the young in America, while also addressing the need for creative approaches to learning and enthusiastic adult involvement in improving students' academic outcomes? The benefits of a learning partnership between adult and child are clear, yet parents and teachers feel pressured for time and resources. I know from experience that parents, in particular, are in search of meaningful ground on which to connect with the educational experiences of their children, and they need reliable, non-commercial tools for engagement. How can they support the academic development of their children when they do not feel inspired by or up to the challenge of the material? I decided that I wanted to make a book neither too broad for the purposes of review nor too abstruse for the purposes of introduction, one that uses vivid imagery to reduce the cognitive load readers can experience when faced with dense and difficult content.

Making Connections, Making Suggestions

With the help of friends and colleagues, I selected several concepts from classical and modern physics that I thought might be suitable for an audience of beginners, and I presented that word list to a physicist and an artist. Each was given the task of creating a single-page description for every term while I searched for literary passages and fine art that might relate comparatively to the scientific ideas (e.g., Aleksandr I. Solzhenitsyn's *The Gulag Archipelago* and Paolo Uccello's *The Battle of San Romano* for the chapter on "Chaos," and Homer's *The Odyssey* and Jacques-Louis David's *The Death of Marat* for "Vector"). The goal in approaching physics from diversified points of view was to delineate for readers a breadth of possible interpretation. Specifically, I wanted to show that the language of imagery and the language of metaphor can yield where the more established language of science cannot. And, if the meanings conveyed by these languages (that is to say, art and science) are not necessarily equal, their impact can certainly be seen as parallel and proportionate. The suggested readings and questions offered in every chapter should serve to deepen the relevance of the book's content overall, contributing to the ability of these "constellations of meaning" to challenge and to engage the reader. In the end, I hoped to achieve for every physics term something akin to a fossil in the sense of lastingness and to a hieroglyph in the sense of richness of meaning.

Motion: A Sample Chapter

Without exception, the fine art for this book was selected in conjunction with the literary passages. Every work of art was chosen in relationship to an author's words, which were inspired, in turn, by the original physics. In regard to the chapter on "Motion," for example, I thought not of objects, but of music. Recalling the centrality of *Halley's Concerto* in Ayn Rand's *Atlas Shrugged*, I looked through the novel and easily found a passage in which Rand compares musical notes to motion. Although there are many artists whose work could be said to symbolize motion (e.g., Jackson Pollock springs instantly to mind), I was compelled by Rand's writing to envision notes as stages of motion or platforms of motion, as motion events, unique entities without which there would be no whole, not unlike the steps of a staircase. In picturing these "events" and the energy-filled spaces between them, I imagined a score, but not just *any* score: I imagined Beethoven's *Ninth Symphony*. Though I am familiar with this incredible composition, it still moves me deeply to see a page from it at the close of the "Motion" chapter. It is my sincere hope that readers will be similarly affected.

The suggested readings for "Motion" include *In Search of Schrödinger's Cat* by John Gribbin, *The Lives of a Cell* by Lewis Thomas, *A People's History of the United States* by Howard Zinn, and *Science & Music* by Sir James Jeans. It bears mentioning that none of the reviews featured in the *Read About It* sections were solicited, and no money or services were exchanged to guarantee placement of a title. The selection criteria were simply to find, acquire, and briefly describe excellent learning materials from a variety of sources. These have been offered to readers who would like to advance to more specific texts and to those seeking objective advice in making choices for research and/or library-building. The *Talk About It* paragraphs are composed of open-ended questions meant to provoke thoughts and discussion inspired by the featured term that pertain more directly to contemporary culture than to physics. The objective of this section is to encourage readers to actively consider social issues, behaviors, and standard modes of thinking, and to teach by example the prerogative to maintain a "freedom of mind." In "Motion," for example, such considerations center on creativity as motion—such as how is dominant culture the fixed "frame of reference" for the manufacture of art, why is it that people are "moved" to create aesthetic expressions of life experiences, and in what ways do themes of motion and stillness resonate in paintings and music?

Creative Determination

Each of the thirty-nine chapters was constructed in a similar manner. After we selected each literary and artistic work, the reproduction rights were researched, requested, and secured. Obviously, the connections made in this book do not indicate direct links to physics, nor do they indicate intentionality on the part of the artists to refer to scientific theories in their work. The associations made are purely the property of the imagination. Although, on occasion, the fine art resembles the collage art, any similarity between the two is accidental ("Motion" offers a striking example of this coincidence, as does "Antimatter," "Field," "Radiation," and "Work").

The only denial for reproduction rights that we received came from the estate of James Joyce—we were not given permission to use an excerpt from *Ulysses* to begin the chapter on "Refraction." I mention this omission only to illustrate the motives that drove the choices made for this book. By choosing *Ulysses* to depict the idea of refraction, I had hoped to show individually recognizable words combined in a way that made them appear entirely new. I wanted to elucidate a *skewing*—that is a *refraction*—in the "space" between the author's presentation and the reader's expectations. Further, I wanted to acquaint readers with a work of art that, like physics itself, has become emblematic of abstruseness and exclusivity rather than of creative determination. I had hoped that this inclusion might lead to discussions about the importance of intellectual freedom in all forms of expression. Surely, of all the threats people face, the threat to original thinking is one of the greatest. It is particularly grievous that younger generations have been discouraged from fulfilling their right and their *responsibility* to challenge established positions via self-generated outlooks and ideas; sadly, the majority of young people have been absorbed into the machine of conformity.

Accuracy & Narrative Competence

One of the chief lessons I learned during this project has been that there are instances in which rigidity in language is non-negotiable, and others in which inexactness can be enthralling, illuminating, *leading*. Scientists speak carefully because there is much at stake in accuracy, and, indeed, the physics text in this book was meticulously prepared. However, those of us who are not scientists cannot—and perhaps should not—be held to the same standards for precision. We are at liberty to consider and to signify more experimentally than science generally allows. Although the place for exactness in science is not to be disputed, we might ask ourselves whether or not a reliance on specificity *to the exclusion of play* with beginners carries with it a certain degree of risk. Can this tendency extinguish natural interest? Can it result in the suppression of early competencies? Can it contribute to a lasting disinterest in the given subject?

I strongly believe that material enhanced by allusion has the capacity not only to enrich the quality of our impressions but the integrity of our creations. The ability to make bold associations and to tell stories is life-enriching and life-affirming. Such narrative competence has broad applications. It positively affects a range of disciplines from science to art to sports to law. Rita Charon, an internist with a Ph.D. in literature who runs the Program in Narrative Medicine at Columbia University, describes the practice of narrative medicine as "medicine practiced with the narrative competence to recognize, absorb, interpret, and be moved by the stories of illness."[2] The phrase *moved by* begs the question, how is it that people become *unmovable*? Could it be the pervasive sense that there is nothing new, no possibility to achieve a spectrum of equally valuable and original outcomes? Yet, in truth, originality is everywhere. Even if so-called "consumer culture" traffics in obscuring and discouraging opportunities to be original, the power to carve these opportunities out again becomes the domain of the *observer*—the scientist, the artist, the writer, the intellectual, the naturalist, the humanitarian, and, yes, the child and the parent. Every individual featured in this book exemplifies the critical importance of risk-taking and imagination. Some were expected to be innovators in a time without precedent for such innovations (e.g., Archimedes, Eratosthenes, and Copernicus, et al.) while others managed to find a way to think unconventionally in a world teeming with extraordinary heroes and role models (e.g., Newton, Einstein, Feynman).

A More Exacting Wonder

As a writer, I am called upon to translate my sense of the world into something transmittable. After working on this book, I have an improved understanding of the science underlying that sense. As a parent, I am called upon to encourage my children to pursue interests in all areas, including those in which I have limited experience. Thanks to what I've learned from working on this project, I am better prepared to guide the education of my children. Finally, as a citizen, I am called upon to familiarize myself with scientific projects and policies that I and my family may rely upon or choose to reject—this book has increased my awareness immensely. After having devoted more than a year of my life to compiling this material, I cannot say that I've become an expert in physics, but I can say that my wonder has become more *exacting*. This book is marked by the generosity and professionalism of its contributors. I am indebted to Emiliano Sefusatti, a dedicated physicist who brought to the project a deep intelligence and an extraordinary integrity. I have been improved by his serious sense of purpose and his gentle guidance. I am thankful to John Morse for his artistry, patience, and his good will. The earliest draft of the manuscript was prepared by Stephen Markacs and Mary Peters, and I am grateful for their creative contributions. Mary's exquisite sense of play and Stephen's ingenious manner of teaching set the exact right tone for the book. Stephen's influence can still be felt in chapters on "Electricity," "Entropy," "Particle," "Radiation," "Uncertainty," and "Wave," and Mary's influence, charm, and love of learning permeates the book. I am thankful as well to the other dedicated physicists who contributed to the science text—Alvaro Nunez Nikitin, Ph.D., and Gabrijela Zaharijas, Ph.D. Thank you to Christian McGrath for the beautiful retouching work, to Michael Nelson for valuable design advice, to Marion St. Laurent for enthusiastic ideas concerning project direction and tone, and to all of the wonderful interns who worked at Vernacular Press these past years. I thank Hillery Hugg for improving my work and my mind wholly. I am fortunate to have the unconditional support of partners and friends—Christine Vecoli, Matt Colabraro, Jovan Franchetti, Deborah Silva, Marco Roth, Eric Great-Rex, Sally Egbert, Leah Glennon, and David and Amy Koza—who inspire me with their willingness to remain impressionable—*movable*—to the stories of humanity. Finally, thanks to my extraordinary family—my parents; my sister, Penelope; my children, Vee, Emmanuelle, and Rainier; and my husband, James—who sponsors so much of my work and who fills every day with lessons that are most completely original.

Hilary Thayer Hamann, New York City, June 2005

Salviati: [...] io grandemente dubito che Aristotele non sperimentasse mai quanto sia vero che due pietre, una più grave dell'altra dieci volte, lasciate nel medesimo instante cader da un'altezza, v. g., di cento braccia, fusser talmente differenti nelle lor velocità, che all'arrivo della maggior in terra, l'altra si trovasse non avere né anco sceso dieci braccia.

Salviati: [...] I greatly doubt that Aristotle ever tested by experiment whether it be true that two stones, one weighing ten times as much as the other, if allowed to fall, at the same instant, from a height of, say, 100 cubits, would so differ in speed that when the heavier had reached the ground, the other would not have fallen more than 10 cubits.

TWO NEW SCIENCES
GALILEO GALILEI

We cannot know whether Aristotle actually performed the experiment to which Galileo refers here. If he did, Aristotle probably did not bother to measure the initial positions of the stones—or their speeds and masses—with great precision. What we *do* know now is that the two stones would have reached the ground at the same time. But with a statement like the one given above, Galileo set the rules of the game for physics at the time. The key word he uses is "experiment." In claiming that Aristotle was wrong, Galileo does not confine himself to words; rather, he suggests experiments to be the ultimate criterion in proving a theory's validity.

Most significantly, after Galileo's work, it was no longer sufficient to hypothesize about events in the physical world based on intuition. Every scientific statement had to be supported by facts and by measurements of physical quantities. If one claims that a stone 10 times heavier than another falls with a 10 times greater speed—and therefore will reach the ground when the second stone will cover only one tenth of the distance—then such a statement must be tested. Speeds and distances have to be measured. The number predicted by Aristotle's theory (10 cubits) has to be compared with the number measured in the experiment.

We cannot say that science began with Galileo. Scientists, or natural philosophers, as they were called, had played the "game" of science for centuries before him. However, Galileo was the first to set clear rules. Scientific statements could not be "proven" or "disproven" on the basis of abstract arguments alone. Galileo's experimental approach provided for theories an objective test on which everyone, ultimately, would have to agree. The central role in science subsequently assumed by experiments forced the scientific community to define the physical quantities to be measured and to define the mathematical relationships between them. Although physics borrowed words such as "mass," "energy," "force," etc. from common language, the discipline gave these words new scientific definitions. The word "power," for instance, which has a variety of broader meanings, has one precise definition in physics, and that word is used by scientists in a very strict and consistent sense.

Categories—On the Beauty of Physics is not a physics textbook. It does not provide rigorous definitions of physical quantities. Rather, in the space of each single page, it indicates the direction of a term's physical meaning and its associated concepts. The equations that appear in many of the chapters are not fully explained; they appear simply to demonstrate the importance of the relationship between physics and math, and to show that every quantity in physics is defined by an equation. Often, when a physicist considers a physical quantity, he or she is thinking of an equation, not a dictionary definition. In most cases, we confined the descriptions of terms found in "classical physics." Classical physics comprises the physical theories that, at the end of the 19th century, explained how forces affect the motion of bodies (mechanics), described most gravitational and electromagnetic phenomena (Newton's theory of gravitation and Maxwell's electromagnetism), and the nature and transformations of heat (thermodynamics).

At the beginning of the 20th century, classical theories proved to be of limited validity. As experiments and equipment became more sophisticated, previously unobservable phenomena required new explanations. New theories were developed. The subatomic world was described in terms of "quantum mechanics." Einstein's special theory of relativity modified even the basic notions of time and space, while his general theory of relativity provided a radically different theory of gravitation. As a consequence, modern theories in physics enriched and redefined the meaning of common physical quantities such as *velocity*, *mass*, or *energy* by assigning to them unexpected properties. Whenever pertinent, we tried to give the different interpretations of physical concepts in different theories. In most cases, we referred to the classical concepts, which are more directly related to the experience of everyday life and, thus, easier for beginners to understand. Clearly, it takes many years to become fluent in the (mathematical) language of physics. Our hope is to instill interest in the exactness of this language and to provide the justification for such precision.

Emiliano Sefusatti, Ph.D., New York City, June 2005

ACCELERATION 01

The term "communication" has had an extensive use in connection with roads and bridges, sea routes, rivers, and canals, even before it became transformed into "information movement" in the electric age. Perhaps there is no more suitable way of defining the character of the electric age than by first studying the rise of the idea of transportation as communication, and then the transition of the idea from transport to information by means of electricity.

When information moves at the speed of signals in the central nervous system, man is confronted with the obsolescence of all earlier forms of acceleration, such as road and rail. What emerges is a total field of inclusive awareness. The old patterns of psychic and social adjustment become irrelevant.

Metropolitan space is equally irrelevant for the telephone, the telegraph, the radio, and television. What the town planners call "the human scale" in discussing ideal urban spaces is equally unrelated to these electric forms. Our electric extensions of ourselves simply by-pass space and time, and create problems of human involvement and organization for which there is no precedent. We may yet yearn for the simple days of the automobile and the superhighway.

UNDERSTANDING MEDIA: THE EXTENSIONS OF MAN
MARSHALL MCLUHAN

From: **AC-CEL-ER-A-TION** *n* (1531)
1 : THE ACT OR PROCESS OF ACCELERATING : THE STATE OF BEING ACCELERATED
2 : THE RATE OF CHANGE OF VELOCITY WITH RESPECT TO TIME; *BROADLY* : CHANGE OF VELOCITY

ACCELERATION

Acceleration is a measure of the rate of change in an object's *velocity*. Since the velocity of an object represents its *speed* and its *direction of motion*, an object experiences an acceleration every time its speed increases or decreases and every time its direction of motion changes. Velocity is a *vector*, something with both magnitude and direction, and so acceleration is represented by a vector that points *in the direction in which the velocity is changing* with magnitude given by the rate of change.

A car's gas pedal is referred to as the "accelerator." More pressure to the pedal results in an increase in the car's velocity—an acceleration. If a car moving along a straight line has an acceleration of 5 meters per second squared (that is, per second per second) the speed of the car increases every second by 5 meters per second. So, if at one moment it has a speed of 20 meters per second, one second later it will move at 25 meters per second. The brake also causes acceleration because it changes the speed of the car—*in the direction opposite to the direction of the motion*—causing the car to slow down. This is called *deceleration*, though the word is seldom used in physics. Deceleration is actually no more than negative acceleration. A car is accelerated not only when its speed increases or decreases, but when it turns. Turning the wheel causes the car to accelerate even if the speed remains constant because turning represents a change in the car's direction of motion. Acceleration is a change in velocity, not just speed. For an object moving along a circle at a constant speed, acceleration is a vector that points to the center of the circle, a phenomenon in physics that is referred to as *centripetal acceleration*. Such acceleration is always perpendicular to the trajectory.

We need vectors to describe quantities such as velocity or acceleration because, in a three-dimensional world, objects can move in different directions. In a three-dimensional space, a vector has three components: we can have acceleration in the forward-backward, left-right, or up-down direction. In formulas, a vector is indicated by a symbol with an arrow on top. This is a compact way to express that a physical quantity has several components.

***See:* Motion, Position, Vector, Velocity**

$$\vec{a} = \frac{\Delta\vec{v}}{\Delta t}$$

[Eq. 1] ***The average acceleration ($\vec{a}$) of a body in a time Δt is given by the ratio of the change in velocity ($\Delta\vec{v} = \vec{v}_{final} - \vec{v}_{initial}$) to the time elapsed ($\Delta t = t_{final} - t_{initial}$). The symbol Δ in front of a physical quantity often indicates how much such quantity changes, a difference between final and initial values.***

THINK ABOUT IT: ***Forces & Acceleration***

According to *Sir Isaac Newton's* [1642-1727] First Law of Motion, *inertia* represents an object's tendency to keep its velocity constant—in speed and direction—unless external forces act upon it. If an object is moving, it will keep moving: no force is necessary to keep it in motion. If an acceleration is detected, some force must be acting on the body. For instance, if we see a ball slowing down—decelerating—it is due to *air resistance* or *friction*, forces that always act in a direction opposite to the direction of motion. Acceleration, not just motion, is always associated with an external force.

READ ABOUT IT: ***Comics, Fast Technology, & I.Q.***

The Cartoon Guide to Physics [1991] by Larry Gonick and Art Huffman — The black-and-white, comic-strip style illustrations in this informative book are edgy and fun. Physics concepts and basic formulas are delivered in an appealing manner.

Faster: The Acceleration of Just About Everything [1999] by James Gleick — How has modern technology changed our lives? Gleick, who was twice nominated for the National Book Award, discusses the socio-cultural impact of time-saving inventions such as "Door Close" buttons on elevators and the use of "speed" drugs like amphetamines.

Flowers for Algernon [1959] by Daniel Keyes — This Nebula Award-winning novel is a first-person account of Charlie, who has an unusually low I.Q. After a medical procedure, Charlie becomes more intelligent than those who seek to help him. Acceleration is approached figuratively with the juxtaposition of Charlie's rapidly changing mental abilities and internal state with external reality.

TALK ABOUT IT: ***Beyond Physics—Acceleration & Control***

Does an increasing number of avenues for information (cable television, the Internet, mobile phones) correspond to increasing amounts of information? ... Is "the news" really new? ... How does accelerated living contribute to feelings of being accepted/rejected, satisfied/frustrated, exceptional/mediocre? ... The faster events occur, the less time we have to consider them ... How does shopping give some people a sense of control over their lives? ... Are the needs of a community met by the media and the marketplace, or do these entities aim to alter perceptions so that "needs" can be defined and controlled? ... Can accelerated living go on "uninterrupted in speed and direction," or will some "external force" bring about change? ...

***Nighthawks* [1942] by Edward Hopper**
Oil on canvas
The Art Institute of Chicago, Chicago

This poetic painting captures the disenfranchisement that was as much a consequence of World War II as it was of the modern age and the resultant *acceleration* of living. Whether these characters are taking refuge from loneliness in the mutual anonymity of an all-night diner, are second-shift workers, or simply prefer to occupy the night, one gets the sense of an unnaturally extended day. The diner is a familiar icon in American art, literature, and cinema. The implication of dining at a counter is that there is no obligation to be formal or polite: a diner is a convenient meeting place for the displaced and an equally convenient means by which artists can comment on displacement. Interestingly, any evidence of the characters' inner despair is undermined by Hopper's tailored and spacious environment.

***Understanding Media: The Extensions of Man* [1964]**
by Marshall McLuhan

In this enormously readable collection of essays, the author discusses the potential impact of *acceleration*—particularly of information—on societies. McLuhan was a thinker of influence and vision. In chapters such as "Clocks," "Movies," "Weapons," and the renowned "The Medium is the Message," McLuhan writes of media, technology, and art as extensions of human awareness. In "Games," he states that "science has become quite self-conscious about the play element in its endless experiments with models of situations otherwise unobservable." Stressing the importance of play as a means by which people can separate imagined consequences from real ones, McLuhan states that sporting events are society's way of "talking to itself."[1]

ANGLE 02

Events are made into a story by the suppression or subordination of certain of them and the highlighting of others, by characterization, motific repetition, variation of tone and point of view, alternative descriptive strategies, and the like—in short, all of the techniques that we would normally expect to find in the emplotment of a novel or a play. For example, no historical event is intrinsically tragic; it can only be conceived as such from a particular point of view or from within the context of a structured set of events of which it is an element enjoying a privileged place. For in history what is tragic from one perspective is common from another.

Interpretation thus enters into historiography in at least three ways: aesthetically (in the choice of a narrative strategy), epistemologically (in the choice of an explanatory paradigm), and ethically (in the choice of a strategy by which the ideological implications of a given representation can be drawn for the comprehension of current social problems).

TROPICS OF DISCOURSE: ESSAYS IN CULTURAL CRITICISM
HAYDEN WHITE

From: **AN-GLE** *n* (14C)
2 A : THE FIGURE FORMED BY TWO LINES EXTENDING FROM THE SAME POINT B : A MEASURE OF AN ANGLE OR THE AMOUNT OF TURNING NECESSARY TO BRING ONE LINE OR PLANE INTO COINCIDENCE WITH OR PARALLEL TO ANOTHER 3 A : THE PRECISE VIEWPOINT FROM WHICH SOMETHING IS OBSERVED OR CONSIDERED; *ALSO* : THE ASPECT SEEN FROM SUCH AN ANGLE B : (1) : A SPECIAL APPROACH, POINT OF ATTACK, OR TECHNIQUE FOR ACCOMPLISHING AN OBJECTIVE

ANGLE

AN *ANGLE* IS SOMETIMES DEFINED as the portion of space between two lines that intersect at a common point called *vertex*. We can also think of an angle as the measurement of the amount of rotation that is needed to make one line coincide with the other. Unlike linear measurements, which give the length of an object in units such as *meters* or *inches*, angular measurements describe a difference in direction, and they are usually expressed in *degrees*. An angle that describes a full rotation measures 360 degrees. Ancient Babylonians were the first to use degrees to measure angles, and it is possible that the 360 degrees of a circle derive from the 360 days of the Babylonian year. In one year, the Earth makes a complete turn, or *revolution*, around the Sun.

In drawing the map of a city, all distances are reduced or scaled down to fit on the page, while the angles made by the streets remain the same. The careful reproduction of angles allows us to maintain the proportions between distances. If we know the internal angles of a triangle, we know the proportions between its sides, and by measuring just one side, we can calculate the other two. Angles are easier to measure than distances, a fact that has been known since antiquity. When periodic flooding washed away stone property markers along the Nile River, surveyors re-measured the borders using angles in order to resolve property disputes between neighbors.

The method of using angles to determine the position of objects in the distance is called *triangulation*. We can determine the height of an apartment building by walking a certain distance away from its base and measuring the angle between the horizon and the building's top. We can then derive all the angles of the imaginary triangle as defined by our location, the base of the building, and its top. Our distance from the building's base—*one side of the triangle*—will help us figure the other two sides, including the height, which is obtained without ever leaving the ground. In land mapping, just as in elementary geometry, it is useful to divide a complicated figure into triangles, since the relations between the sides and angles of a triangle are particularly simple to understand.

***See:* Measurement, Position**

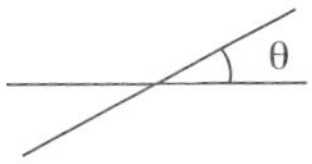

θ

[Ill. 1] ***Whenever two nonparallel lines in a plane meet, they form an angle.***

THINK ABOUT IT: *Measuring the Earth*

Eratosthenes [276-194 B.C.E.] calculated the size of the Earth using very simple geometry. He assumed that since the Sun is very far away, its rays must reach the Earth parallel to each other. He noticed that since the Sun is always in the same position in the sky on a given day of the year at a given hour, its rays must always hit the Earth with the same *inclination*. If, in fact, the Earth were flat, this inclination would have to be the same in every place on Earth. But this was not the case. Eratosthenes measured different *angles of inclination* by measuring the shadows at noon in Alexandria and in Siene, a city in southern Egypt, on the day of the summer solstice. On a curved surface, parallel rays hit different points at different angles. By estimating the distance between the two cities, he was able to derive the circumference of the Earth to be about 250,000 "stadia," corresponding to a distance of about 45,000 KM, very close to the actual value, which is 40,075 KM.

READ ABOUT IT: *Tally Sticks, Geometry, & Snowflakes*

The Mathematical Traveler: Exploring the Grand History of Numbers [1994] by Calvin Clawson — In this version of the history and theory of numbers, Clawson describes ancient counting tools (e.g., tally sticks made of wood and wolf bones), Babylonian math, "Counting in Other Species," "The Story of Pi," and the infinite.

Physics Demystified: A Self-Teaching Guide [2002] by Stan Gibilisco — The simple layout of this comprehensive review of math and physics makes the subject matter approachable. Each chapter, such as "The Basics of Geometry," includes tests (and answers!).

What Shape Is a Snowflake?: Magical Numbers in Nature [2001] by Ian Stewart — Observing patterns in nature is a fine way to explore science and math. Stewart's lovely book covers order and disorder, spots and stripes, spiral growth, turbulence, domes, and viruses.

TALK ABOUT IT: *Beyond Physics—Angles & Interpretation*

In science and math, angles are "the figures formed by two lines extending from the same point" ... Consider converging "lines" of interpretation and the spaces—or the angles—created between and around them ... How does an opinion form a "line" once it has been expressed? ... Consider the way that every acute angle refers to a companion obtuse angle ... How might a meaningful conversation between friends include corresponding "angles of expression"? ...

Mont Sainte-Victoire **[1897-1898] by Paul Cézanne**
Oil on canvas
The State Hermitage Museum, St. Petersburg

The *angles* in this painting are subtle but numerous—the inward-moving layers of earth, the implied avenues in rock, the diverse planes from which the observer might gaze. They dip and rise, sometimes supple, sometimes firm—always contributing to a dynamic sense of place. Unlike that of his Impressionist predecessors—Monet, Degas, and Renoir, among others—Cézanne's lyricism is rooted in geometric fullness. He once advised fellow artist Emile Bernard (See: *Work*) to "treat nature by means of the cylinder, the sphere, the cone ... so that each side of an object or a plane is directed toward a central point."[2] Cézanne is said to have greatly admired this particular subject: he painted Mont Sainte-Victoire scores of times.

Tropics of Discourse: Essays in Cultural Criticism **[1978] by Hayden White**

Is it really possible for a written history to be unaffected by the perspective—i.e., the *angle*—of the author and the intellectual, social, and political systems of the author's time? In this philosophical study of the means and methods of recording history, White questions whether historians or writers (or people in general) can ever offer any perspective more rigorous or objective than "tropics," that is to say, general lines, of discourse. The points made by the author concerning the subjectivity of interpretation provide excellent grounds for cooperative discussion.

ANGULAR VELOCITY 03

Then he saw a blade that boded well,
a sword in her armoury, an ancient heirloom
from the days of the giants, an ideal weapon,
one that any warrior would envy,
but so huge and heavy of itself
only Beowulf could wield it in a battle.
So the Shieldings' hero, hard-pressed and enraged,
took a firm hold of the hilt and swung
the blade in an arc, a resolute blow
that bit deep into her neck-bone
and severed it entirely, toppling the doomed
house of her flesh; she fell to the floor.
The sword dripped blood, the swordsman was elated.

BEOWULF
BY UNKNOWN
TRANSLATED BY SEAMUS HEANEY

From: **AN-GU-LAR VE-LOC-I-TY** *n* (1819)
1 : THE RATE OF ROTATION AROUND AN AXIS USUALLY EXPRESSED IN RADIANS OR IN REVOLUTIONS PER SECOND OR PER MINUTE

ANGULAR VELOCITY

ANGULAR VELOCITY MEASURES HOW FAST an object rotates, and it gives the direction of the axis of rotation. For this reason, we need a vector to describe it. While speed is a distance per unit time, the magnitude of the angular velocity is defined as a *change in angle per unit time*. For example, when two children are riding on a merry-go-round at different distances from the axis (one is near the center, the other is near the rim), their velocities are different—the one near the rim is going faster—but their angular velocity is the same: *they span the same angle in the same time*. Sometimes it is enough to give the *number of turns per unit time* to indicate how frequently objects return to the initial position. The number of turns per unit time is referred to as *frequency*.

A simple example of angular velocity is given by an ordinary wristwatch. The second hand makes a complete circle in 60 seconds. To determine the number of *degrees per second*, we need only to divide 360 degrees by 60 seconds: 360/60 = 6 degrees per second. This is the same as saying *360 degrees per minute*. In each case, we have an angle measurement per some unit of time. It is important to know that angular velocity is a vector: it has magnitude and direction. If we were to draw the vector representing the angular velocity of the watch's hand, we would draw an arrow along the axis of rotation, pointing *into the face of the watch*. The fact that this vector points into the face is relevant because it indicates, according to convention, that the hands are moving *clockwise*.

We need more than one angular velocity to describe a spinning top. Not only does the top rotate on its own axis, but the axis itself rotates about the imaginary vertical line that passes through the point where the top touches the ground. This second rotation is called *precession*. Thus, we need at least two angular velocities to describe the motion of a top: the vector along the axis of the top itself, and the vector along the imaginary vertical line perpendicular to the ground. *The first rotates about the second*. However, this description is just an approximation. Things become more complicated when the top slows down and begins to wobble.

***See:* Angle, Measurement, Position, Vector, Velocity**

$$w = \frac{\Delta\theta}{\Delta t}$$

[Eq. 2] ***The magnitude of the average angular velocity (w) of a rotating body in an interval of time Δt is given by the ratio of the change in angle $\Delta\theta$ to the time elapsed.***

THINK ABOUT IT: *Planetary Angular Velocity*

Every planet rotates around its own axis and along an *elliptical orbit* around the Sun. The angular velocity corresponding to the Earth's rotation on its own axis is given by 360 degrees divided by 24 hours—that is to say, 15 degrees per hour. The Earth orbits the Sun in 365 days with an angular velocity of about 1 degree per day, or 0.041 degrees per hour. The *solar system*—the Sun and all the planets orbiting it—is, in turn, rotating around the center of our galaxy, the Milky Way. The period of such rotation is about 240 million years. That rate would correspond to a very small angular velocity, less than a billionth of a degree per hour. On the other hand, the Sun's orbital velocity in the galaxy is approximately 225 kilometers per second, or about 500,000 miles per hour: light is "only" 1,000 times faster!

READ ABOUT IT: *The Milky Way & The Big Bang*

Coming of Age in the Milky Way [1988] by Timothy Ferris — In this richly detailed book, Ferris discusses the evolution of human ideas of time and space in language that is lively and accessible. Drawings, famous quotes, and formulas are interspersed throughout.

The Concise Science Encyclopedia [2000] by Kingfisher Publications — Whether this book is used for school projects or as a starting point for early inquiry, it will be much appreciated. Chapters such as "Momentum," "Galaxies," and "Catalysts" are colorful and clear.

Understanding Cosmology [2002] by *Scientific American* — This compilation of essays from *Scientific American* contains some surprising speculations about the origin, structure, and nature of the universe. The discussions of the Big Bang, antigravity, quantum cosmology, and the fate of life are interesting and informative.

TALK ABOUT IT: *Beyond Physics—Angular Velocity & Making Things New*

Galaxies, planets, electrons, tornadoes, whirlpools, seasons ... What is the significance of a cycle? ... How does one rotation or cycle provide energy for the next? ... Think of a rotation or a cycle as a source of predictability or stability (time and the seasons) ... As a form of mystery (moving without stopping into an unknown future) ... As a sign of ongoingness (birth, life, and death) ... As a source of energy renewal (crop rotation) ... What does it mean to "come full circle," "make a round trip," "stir things up," "fall head over heels"? ... When is a rebellion considered a "revolution"? ...

***Equivalent* [1929] by Alfred Stieglitz**
Gelatin silver print
George Eastman House, Rochester, N.Y.

Though we all observe the same universe, our perception is influenced by our own interests and expertise. What an artist might admire for its riveting suggestive power, a scientist might distill to measurement and formula. Stieglitz, leader of the American school of photographers and husband of artist Georgia O'Keeffe (See: *Radiation*), was probably not thinking of *angular velocity* when he took this photograph but rather of the mystery and majesty of the cosmos. Stieglitz dedicated his life to proving that photography is high art. He formed "Photo-Secession," a coalition of like-minded photographers determined to separate art from commerce, and he published their work in the periodical, *Camera Work* [1903-1917].

***Beowulf* [7th-10th century] by Unknown**
Translated by Seamus Heaney

This epic Old English poem achieves brilliant new life in Heaney's translation [2000], which presents the remarkably beautiful ancient text on the left-hand pages. The arc of the battle sword wielded by the heroic Scandinavian prince against the monster Grendel's even more monstrous mother is suggestive of *angular velocity*. J. R. R. Tolkien called *Beowulf* a work of "creative intuition and conscious structuring,"[3] to demonstrate the poem's effect on modern literary traditions.

ANTIMATTER 04

The universe is made of both particles and anti-particles. Our part of it, however, is made almost entirely of regular particles which combine into regular atoms to make regular molecules which make regular matter which is what we are made of. Physicists speculate that in other parts of the universe anti-particles combine into anti-atoms to make anti-molecules which make anti-matter which is what anti-people would be made of. There are no anti-people in our part of the universe because, if there were, they all long since have disappeared in a flash of light.

THE DANCING WU LI MASTERS
GARY ZUKAV

Yet across the gulf of space, minds that are to our minds as ours are to those of the beasts that perish, intellects vast and cool and unsympathetic, regarded this earth with envious eyes, and slowly and surely drew their plans against us. And early in the twentieth century came the great disillusionment.

THE WAR OF THE WORLDS
H. G. WELLS

From: **AN-TI-MAT-TER** *n* (1950)
1 : MATTER COMPOSED OF ANTIPARTICLES

From: **AN-TI-PAR-TI-CLE** *n* (1934)
1 : A SUBATOMIC PARTICLE IDENTICAL TO ANOTHER SUBATOMIC PARTICLE IN MASS BUT OPPOSITE TO IT IN ELECTRIC AND MAGNETIC PROPERTIES (AS SIGN OF CHARGE) THAT WHEN BROUGHT TOGETHER WITH ITS COUNTERPART PRODUCES MUTUAL ANNIHILATION

ANTI.MATTER

THE PARTICLES THAT MAKE UP our universe are identified by a set of numbers called *quantum numbers*. The quantum number that people are most familiar with is the *electrical charge*.

Every particle has an electrical charge; for instance, a proton has a charge of +1 and an electron of -1. Other particles might have no charge, in which case we say that their charge is 0—which is still a number. The electrical charge indicates how the particle will interact with the electromagnetic field. In order to describe the behavior of electrons and their interactions, physicist *Paul A. M. Dirac* [1902-1984] postulated the existence of the *positron*. A positron is a particle with the same mass as an electron but with an opposite charge. We now know that for every particle in nature, there is an *antiparticle* with the same mass, but an opposite charge. This does not mean that for every electron in the universe there are as many positrons existing somewhere else. It only means that if we observe a particle of a given kind, its antiparticle exists—*in principle*—and can be created in a lab. Though no antiparticle had been observed at that time, Dirac's theory required their existence. Shortly after his prediction, electron-positron pairs were created in experiments.

When a particle and its antiparticle interact, they might *annihilate* each other. The word *annihilation* literally means "turn to nothing." The annihilation of an electron and a positron actually results in the production of two *photons*. A photon is a particle of light with no mass and zero electrical charge; thus, the mass of the two particles disappears. However, the total energy is *conserved* in the process—*it does not change*. The mass that disappears corresponds to the energy transferred to the photons. Also, the total electrical charge is conserved. The combined charge of an electron and a positron is zero because one is the opposite of the other, and after the annihilation, *the total charge is still zero*—zero being the charge of the individual photons. The simplest atom in nature is the *hydrogen atom*, which is comprised of a proton in a nucleus and an electron. One can easily imagine an anti-proton and a positron forming an anti-hydrogen atom. Such an atom would then be the simplest atom of *antimatter*.

***See:* Energy, Light, Mass, Particle**

THINK ABOUT IT: *Quantum Mechanics*

The branch of physics that studies how forces affect the motion of bodies is called *mechanics*. *Classical mechanics*, based on Newton's laws of dynamics, turned out to be extremely successful in describing the motion of baseballs as well as the motion of planets and galaxies. But when applied to motion on an atomic scale—for instance, the motion of electrons in an atom—Newton's laws failed to correctly describe what was observed by physicists. A new mechanics was then developed: *quantum mechanics*. The theory of quantum mechanics is not restricted to the behavior of elementary particles: it can also describe the motion of a baseball. But for larger, macroscopic objects, Newton's laws of mechanics are still valid and easier to apply.

READ ABOUT IT: *Particle Accelerators & String Theory*

Antimatter: The Ultimate Mirror [2000] by Gordon Fraser — This Swiss physicist tracks the developments in the study of the universe, preceding and including fellow physicist Paul Dirac's early theories about antimatter, Einstein's contributions equating matter with energy, and recent experiments being conducted in particle accelerators.

The Elegant Universe [1999] by Brian Greene — Greene, a Columbia University professor of mathematical physics, describes the basics as well as some complicated ideas involving black holes, the importance of string theory (and the nine space dimensions it requires!), and the inability of science to explain everything.

Nova: The Elegant Universe [2003] directed by Joseph McMaster — This fascinating science documentary is based on Greene's book, mentioned above. Greene acts as host. Produced by WGBH, Boston.

TALK ABOUT IT: *Beyond Physics—Antimatter & Absence*

Think of antimatter as a "defining absence" (e.g., the defining absence of a cup is the space into which liquid is poured, or windows are useful by virtue of the missing wall) ... In some sports, the emptiness defines the terms of the game—the basketball net, the golf hole, the hockey goal ... In art, there is a paradigm of absence—the absence prior to creation (the nothingness filled by the somethingness of art), the absence inherent in representation (the artwork is not *the thing depicted), and the importance of the negative space to the rendering of the positive "subject" of the work ... What is present in/absent from film? Music? Literature? ... If the body is "matter," what is its "antimatter"? ... The soul within? ... The aura beyond? ...*

***Venus* [1952] by Henri Matisse**
Paper collage on canvas
National Gallery of Art, Washington, D.C.

In art and science, what *can* be seen is often best understood in conjunction with what *cannot* be seen. The wonder of Matisse's *Venus* cut-out—made when he was 82 years old and confined to bed—can be found in the juxtaposition of positive and negative space, and in the joyous play with absence and presence (or, that is, of matter and *antimatter*!). The blocks of color are inscribed with secret dimension—all that Venus is *not*, by implication, is all that we are.

***The Dancing Wu Li Masters* [1979] by Gary Zukav**

No science background is needed for a book that is so generously devoted to the growth of its readers and so charmingly enamored of its subject. Zukav is clearly having a good time as he describes *antimatter* with images of anti-atoms, anti-molecules, and anti-people!

***The War of the Worlds* [1898] by H. G. Wells**

This science-fiction classic describing an alien invasion of Earth was adapted so believably for radio in 1938 by Orson Welles (no relation to H. G.) that the mock broadcast caused hysteria. H. G. Wells was a social utopian and a visionary author whose approach to science was cautionary and inventive. He speculated prophetically about aerial warfare and radioactive decay. Here, he seems to hint at *antimatter*.

CHAOS 05

I was walking along the highway among wrecked and overturned German automobiles, and a wealth of booty lay scattered everywhere. German cart horses wandered aimlessly in and out of a shallow depression where wagons and automobiles that had gotten stuck were buried in the mud, and bonfires of booty were smoking away. Then I heard a cry for help: "Mr. Captain! Mr. Captain!" A prisoner on foot in German britches was crying out to me in pure Russian. He was naked from the waist up, and his face, chest, shoulders, and back were all bloody, while a sergeant osobist, a Security man, seated on a horse, drove him forward with a whip, pushing him with his horse. He kept lashing that naked back up and down with the whip, without letting him turn around, without letting him ask for help. He drove him along, beating and beating him, raising new crimson welts on his skin.

Any officer, possessing any authority, in any army on earth ought to have stopped that senseless torture. In any army on earth, yes, but in ours? Given our fierce and uncompromising method of dividing mankind? (If you are not with us, if you are not our own, etc. then you deserve nothing but contempt and annihilation.) So I was afraid to defend the Vlasov man against the osobist. I said nothing and I did nothing. I passed him by as if I could not hear him ... so that I myself would not be infected by that universally recognized plague. (What if the Vlasov man was indeed some kind of supervillain? Or maybe the osobist would think something was wrong with me? And then?) Or, putting it more simply for anyone who knows anything about the situation in the Soviet Army at that time: would that osobist have paid any attention to an army captain? This picture will remain etched in my mind forever. This, after all, is a symbol of the Archipelago.

THE GULAG ARCHIPELAGO 1918-1956
ALEKSANDR I. SOLZHENITSYN
TRANSLATED BY THOMAS P. WHITNEY

From: **CHA-OS** *n* (15C)
1 *OBS* : CHASM, ABYSS 2 A : A STATE OF THINGS IN WHICH CHANCE IS SUPREME; *ESP* : THE CONFUSED UNORGANIZED STATE OF PRIMORDIAL MATTER BEFORE THE CREATION OF DISTINCT FORMS—COMPARE COSMOS B : THE INHERENT UNPREDICTABILITY IN THE BEHAVIOR OF A NATURAL SYSTEM (AS THE ATMOSPHERE, BOILING WATER, OR THE BEATING HEART) 3 A : A STATE OF UTTER CONFUSION

CHAOS

PHYSICISTS OFTEN USE THE WORD *chaos* to describe a system whose behavior appears unpredictable, although the basic laws that govern it are known. The term is usually applied to systems whose final state is extremely sensitive to initial conditions.

For instance, we know exactly how air molecules in the atmosphere interact with each other, but in order for us to predict the weather with adequate precision, we would probably have to know the position and velocity of *every* particle in the atmosphere at a given initial time—that is, the "initial conditions." Even if this were possible, it is unlikely that we would be able to analyze the huge amount of data that would be generated. Besides, it would hardly make sense to make months of calculations to predict tomorrow's weather. In general, the behavior of some chaotic systems could be predicted if we could measure with the required accuracy their *initial configuration*. We call these systems *deterministic* because their behavior can be determined, in principle, by preceding events and natural laws.

Chaotic systems do not necessarily involve a large number of particles. Three masses that are moving under the influence of their mutual gravitational attraction form a chaotic system. If there are only two masses, such as the Earth and the Sun, we can describe their motion as one that moves along a certain orbit, and we can write an equation for that orbit in a simple form. We also know that if we choose a slightly different initial position for such a mass, its orbit will be only slightly different from the previous one. In the two-body case, if we have a small uncertainty on the position of the bodies at the initial time, such uncertainty will stay virtually the same. But in a three-body case, if we have a small uncertainty on the initial positions, the uncertainty at successive times will grow very quickly. The behavior of the system will be hard to predict. In the case of three bodies, we are still able to calculate their trajectories, but we have to do it by successive approximations: we cannot find simple formulas. What is most important is that a small difference in the initial position of one of the three masses could result, after a relatively short time, in completely different trajectories for the three bodies.

See:* *Entropy, Measurement, Particle, Position, Uncertainty, Velocity

THINK ABOUT IT: ***Chaos & Fractals***

Fractals are the complex images often associated with chaos. Fractals fill space with a repetitive aesthetic: one can enlarge any part of a fractal and find patterns that re-occur on smaller and smaller scales. In this sense, fractals are highly ordered systems. A fractal can be created on a computer by using a simple rule, called an *algorithm*, and repeating it. Sometimes, if the parameters that control the algorithm are changed—by just a little—the resulting image is extremely messy, something that we would commonly think of as "chaos." Mathematicians and physicists think that some chaotic systems might be explained by algorithms as simple as the ones on which fractals are based (not the weather, though!).

READ ABOUT IT: ***Chaos Theory & Patterns in Nature***

Chaos: Making a New Science [1988] by James Gleick — Gleick's thorough and thrilling account of chaos theory, complete with photographs and diagrams, was nominated for the National Book Award and the Pulitzer Prize. *Chaos* is intended for general audiences.

The Fractal Geometry of Nature [1977] by Benoit Mandelbrot — Mandelbrot coined the term "fractal" to describe a "family of shapes" involving chance and statistical irregularities. Fractals are patterns that repeat in different scales within the same self-similar object. He describes the inability of geometry to describe "a cloud, a mountain, a coastline, a tree." This book is an excellent primary resource.

Fractals: The Patterns of Chaos [1992] by John Briggs — This book reads like a large-scale magazine—attractive and approachable. The photos are colorful and slick, and the text is formatted into brief passages. In it, Briggs explains how to make fractals on home computers.

TALK ABOUT IT: ***Beyond Physics—Chaos & War***

Apply the concept of chaos to war (i.e., the "basic laws that govern" war are known, but the outcome is "unpredictable" and "sensitive to initial conditions") ... War involves numerous lines of action—chances for communication/miscommunication—and, thus, many opportunities to trigger events that add to unpredictability ... In peace, there is an emphasis on the private over the public, on accumulation of individual resources, on containment in general ... How does hyper-connectivity via the Internet, mobile phones, and globalization forge ever-increasing "lines of action" and new chances for chaos via communication/miscommunication, anonymity, etc.? ...

***The Battle of San Romano* [circa 1450-1455]**
by Paolo Uccello
Egg tempera and silver foil on wood panel
The National Gallery, London

War begins as a relatively simple idea (defense or offense) and gets shaped into a relatively careful formula (a battle plan); however, once it begins, it falls prey to myriad natural factors that render the outcome entirely unpredictable. Florentine artist Uccello was among the Early Renaissance painters intrigued by the nascent science of linear perspective, and, according to Janson's *History of Art*, "he superimposed this technique on his earlier style like a straitjacket."[4] In this incredible painting, the *chaos* of battle is palpable—the obvious confusion of the subjects, the complicated patterns created by the weapons, the surreal flattening of distant space, the unsettling use of angles, the bizarre palette, the meticulous attention to detail.

***The Gulag Archipelago* [1958] by Aleksandr I. Solzhenitsyn**
Translated by Thomas Whitney

Solzhenitsyn's years in Soviet prison camp [1945-1953] form the basis of what *Time* magazine called the "Best Non-Fiction Book of the 20th Century." Solzhenitsyn draws material from his own experiences as well as from those of 227 other witnesses. In this particular excerpt, he offers a painfully honest description of his having ignored the desperate pleas of a demoralized German officer, thus implicating himself in the *chaos* and collapse of reason that characterized the Soviet post-World War II era, as if to say that his own personal apathy—like the flapping of an insect's wings—played its part in history. This dense book can be enjoyed in its entirety by a single reader, or passages can be read aloud by members of small reading groups.

COLOR 06

You will see that I am not afraid of a bright green or a soft blue, and the thousands of different greys, for there is scarcely any colour that is not grey—red-grey, yellow-grey, green-grey, blue-grey. As far as I understand it, we of course perfectly agree about black in nature. Absolute black does not really exist. But like white, it is present in almost every colour, and forms the greys—different in tone and strength; so that in nature one really sees nothing else but those tones or shades. There are but three fundamental colours, red, yellow, and blue; "composites" are orange, green, and purple. By adding black and some white one gets the endless varieties of greys; to say, for instance, how many green-greys there are is impossible. But the whole chemistry of colours is not more complicated than those few simple rules. And to have a clear notion of this is worth more than seventy different colours of paint, as, with those three principle colours and black and white, one can make more than seventy tones and varieties. The colourist is he who, seeing a colour in nature, knows at once how to analyze it, and to say, for instance, that green-grey is yellow with black and hardly any blue. In other words, he who knows how to find the greys of nature on his palette.

DEAR THEO:
THE AUTOBIOGRAPHY OF VINCENT VAN GOGH
[FROM A LETTER DATED DECEMBER 1881]
VINCENT VAN GOGH
EDITED BY IRVING STONE

From: **COL-OR** *n* (13C)
1 A : A PHENOMENON OF LIGHT (AS RED, BROWN, PINK, OR GRAY) OR VISUAL PERCEPTION THAT ENABLES ONE TO DIFFERENTIATE OTHERWISE IDENTICAL OBJECTS B : THE ASPECT OF OBJECTS AND LIGHT SOURCES THAT MAY BE DESCRIBED IN TERMS OF HUE, LIGHTNESS, AND SATURATION FOR OBJECTS AND HUE, BRIGHTNESS, AND SATURATION FOR LIGHT SOURCES C : A HUE AS CONTRASTED WITH BLACK, WHITE, OR GRAY

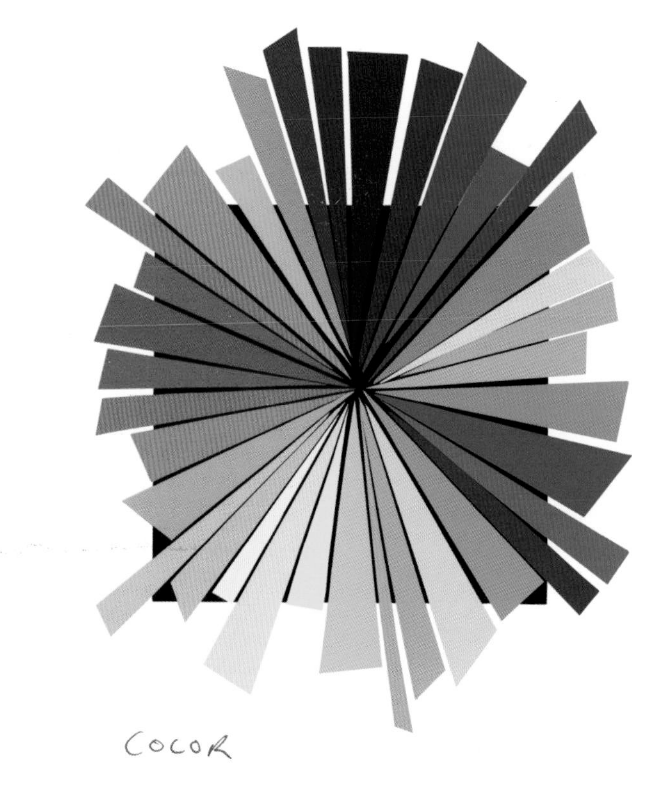
COLOR

Our understanding of the world is enhanced by our ability to see *color*. When light is emitted directly from a source or reflected by an object, it results in a sensation our eyes interpret as *color*.

In physics, light is described as an electromagnetic wave and like any wave, it is characterized by a *frequency*. The human eye can detect only a certain range of frequencies, and different colors correspond to the light of different visible frequencies. Black is not considered to be a true color (actually, it has no hue), but rather the perceived absence of light. White, on the other hand, is a combination of all visible frequencies. This fact is clearly observed when a beam of white light is split into the different colors of the visible spectrum by a prism. The sequence of colors from the lowest to the highest frequencies are red, orange, yellow, green, blue, indigo, and violet (commonly referred to by its mnemonic, ROY G BIV). We call *infrared* the electromagnetic waves of frequencies below the frequency of red light, and *ultraviolet* those with frequencies above the violet one.

Objects can have different colors for many different reasons. In the simplest case, a light source has a certain color when it emits light of a determined frequency, as in the case of a laser beam. Most of the colors we see are due to the way objects absorb light. An object can absorb certain frequencies and reflect others. For instance, a black object will absorb all visible frequencies and reflect none. Microscopically, the absorption of light depends on the electromagnetic properties of the object. Light is an electromagnetic wave; it consists of oscillating electric and magnetic fields, while atoms and molecules are made up of electrically charged particles—protons and electrons. An electric charge placed in an electric field experiences a force. A beam of light then affects the vibrational motion of electrons in atoms and molecules—*the oscillating electric field makes the electrons oscillate*. In this way, electric charges in atoms and molecules can absorb just the energy of the electromagnetic wave, turning it into heat as the vibrational motion is transmitted to other molecules, or, since oscillating charges can in turn create electromagnetic fields, the charges can emit—reflect—the electromagnetic wave.

See: Field, Image, Particle, Radiation, Reflection, Refraction, Wave

THINK ABOUT IT: *Color & Perception*

There is an irony inherent in the naming of colors. We tend to think of a color as being a property of the object; for instance, we might think of a red tomato as being "saturated with redness." However, our perception of the tomato's redness is a result of the fact that the tomato has absorbed all the colors or wavelengths of light other than the red ones. Red wavelengths are "rejected" by the object, they bounce off the surface, and these bouncing wavelengths are perceived by our eyes. Obviously, it is more efficient to call a tomato "red" than to call it "every other color but red."

READ ABOUT IT: *Goethe, Visual Image Processing, & The Origins of Color Names*

Theory of Colours [1840] by Johann Wolfgang von Goethe — Goethe's remarkable (though ultimately incorrect) study of the effects of color can be enjoyed for the beauty of his impressions and for the wonder of his ideas. In the book, which relates color to math, history, and music, Goethe directs his attention to appearances and to meaningful descriptions of appearances, and his efforts encourage readers to do the same. What *is* the best way to describe color?

Vision and Art: The Biology of Seeing [2002] by Margaret Livingstone — Livingstone, a neurophysiologist at Harvard Medical School, writes about the nature of light, the structure of the eye, visual image-processing, color perception, and the significance of primary colors. Every page of this big, beautiful book is rich with detail.

When Blue Meant Yellow: How Colors Got Their Names [1994] by Jeanne Heifetz — This description of the origins of color names is delightful reading. The book draws upon history, legend, and anecdote to provide cross-cultural curiosities about color. For example, according to the author, modern Vietnamese has no word for *blue* and Cantonese has none for *brown*.

TALK ABOUT IT: *Beyond Physics—Color & Identity*

What is the biological advantage to human color perception? ... How does color motivate? ... Naturally-occurring versus manufactured color ... Color and political intention ... Color intensity ... Why are flags color-rich? ... How do national colors encourage a sense of belonging and self-respect? ... When we align ourselves, what is our obligation to respect the rights of others to do the same? ... Can identity be solidified without unfair comparisons being drawn? ...

***Madonna del Granduca* [circa 1505] by Raphael**
Oil on panel
Palazzo Pitti, Florence

Not only is the *color* in this exquisite painting remarkable, but so too is the *rhythm* of the color. High Renaissance artist Raphael creates a definite, almost sculptural, impression of center—the open space between the meeting hearts of mother and child—by leading the eye around the sumptuous reds and blues, past the magnificent golden apricot of the baby's bindings, to the gentle touch of the mother's hand. By delineating the tender core of the painting, Raphael draws the viewer into the intimacy of their embrace.

***Dear Theo: The Autobiography of Vincent van Gogh* [1873-1890] by Vincent van Gogh, Edited by Irving Stone**

This collection of letters written by the artist to his beloved brother, Theo, during the years 1873 to 1890 contains revelatory and heart-wrenching descriptions of the creative process. Though he received no formal training, van Gogh was a magnificent artist and brilliant colorist, and the passages in which he describes the properties and the practical applications of *color* are unforgettable. Readers will become acquainted with the workings of the painter's mind, gaining insight into how much thought, effort, humility, courage, and true inventiveness went into the creation of some of the world's most precious artwork. *Dear Theo* is highly recommended.

ELECTRICITY 07

What gave my Book the more sudden and general Celebrity, was the Success of one of its propos'd Experiments, made by Messrs. Dalibard and Delor, at Marly, for drawing Lightning from the Clouds. This engag'd the public Attention every where. M. Delor, who had an Apparatus for experimental Philosophy, and lectur'd in that Branch of Science, undertook to repeat what he call'd the Philadelphia Experiments, *and after they were performed before the King and Court, all the Curious of Paris flocked to see them. I will not swell this Narrative with an Account of that capital Experiment, nor of the infinite Pleasure I receiv'd in the Success of a similar one I made soon after with a Kite at Philadelphia, as both are to be found in the Histories of Electricity.*

THE AUTOBIOGRAPHY OF BENJAMIN FRANKLIN
BENJAMIN FRANKLIN

From: **ELEC·TRIC·I·TY** *n* (1646)
1 A : A FUNDAMENTAL ENTITY OF NATURE CONSISTING OF NEGATIVE AND POSITIVE KINDS, OBSERVABLE IN THE ATTRACTIONS AND REPULSIONS OF BODIES ELECTRIFIED BY FRICTION AND IN NATURAL PHENOMENA (AS LIGHTNING OR THE AURORA BOREALIS), AND USUALLY UTILIZED IN THE FORM OF ELECTRIC CURRENTS B : ELECTRIC CURRENT OR POWER 2 : A SCIENCE THAT DEALS WITH THE PHENOMENA AND LAWS OF ELECTRICITY 3 : KEEN CONTAGIOUS EXCITEMENT

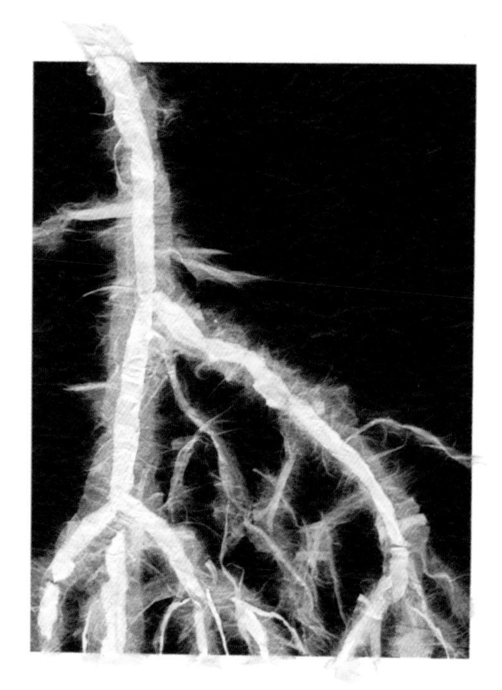

ELECTRICITY

THE *ELECTRICITY* THAT COMES THROUGH wall outlets and makes machines run is actually the motion of particles called *electrons*, and the force responsible for their motion is the *electric force*. In the simple case of two electric charges, the force they exert on one another depends on the distance between them, in exactly the same way that the gravitational force between two masses depends on the distance between the masses. The electric force also depends on the values of the electric charges, but unlike the gravitational force, it can be attractive or repulsive—attractive between a negative and a positive charge, or repulsive when the charges have the same sign. Two charges held at a certain distance apart have an electric potential energy because they have the possibility to move closer together or farther apart, according to their charges.

If two ends of a wire are hooked up to a battery, the electrons in the wire gain potential energy—the possibility to move from one end of the wire to the other. Although the electrons responsible for the electric current will move at a relatively small speed (just a few meters per hour), the "news" that potential energy has been given to the electrons *propagates* quickly through the wire—almost at the speed of light. The *voltage* of the battery is a measure of the difference in potential energy that makes electrons move. The *current* is the measurement of that flow—specifically, the amount of charge that crosses a section of wire in a unit of time. An increase in the voltage results in a greater flow of electrons. *Resistance* is the measurement of the degree to which a given wire will oppose the flow of electrons.

Electricity can flow more easily in some materials than others. Metals are good *conductors* because some electrons can move freely from one atom to another, while in nonconducting materials, such as wood, the electrons are bound to their own atoms and cannot leave them. Current was discovered before electrons were, and while it was clear that something was flowing, the direction of flow was unclear. Thus, according to convention, current runs in the opposite direction of the flow of electrons. A flow of negative charges in one direction is equivalent to a flow of positive ones in the opposite direction.

***See:* Energy, Field, Force, Particle, Power**

THINK ABOUT IT: *Electromagnetism*

The electric force is one aspect of a fundamental force of nature called *electromagnetic force*. The modern description of electromagnetism is given in terms of *fields*. The force between two charges is described as the effect on one charge due to the electric field created by the other charge, and vice versa. Electric charges are not only responsible for an electric field: when the charges move, they determine a magnetic field in the surrounding space. It is impossible to accurately describe the electric force between moving charges without taking into account the magnetic field. Electricity and magnetism are two aspects of the same force.

READ ABOUT IT: *Thunder, Circuits, & Bio-electricity*

All About Lightning [1971] by Martin Uman — Each chapter of Uman's thorough book begins with commonly asked questions, such as "What Should I Do if Caught Outside in a Thunderstorm?" and "Why Did Benjamin Franklin Fly the Kite?" Uman explains why thunder "claps" and what to do in the event of a lightning occurrence.

The Art of Electronics [1980] by Paul Horowitz and Winfield Hill — This sophisticated contemporary classic is intended for those interested in electronic circuit design (analog and digital), microprocessors, and digital electronics. The book is nicely laid out and illustrated, and the authors' teaching method is largely nonmathematical.

The Body Electric [1985] by Robert Becker and Gary Selden — This is an account of Becker's research on the effect of electromagnetic fields on living things—for example, the role of bio-electricity in healing and the dangers of electromagnetic pollution. Becker, an orthopedic surgeon and a leader in the field of regeneration, was nominated for the Nobel Prize—*twice*.

TALK ABOUT IT: *Beyond Physics—Electricity & The Body*

The electric nature of the human body ... How do we produce and distribute personal electricity? ... Can a person's "charge" or "force" be strong or weak, positive or negative? ... What makes someone a "winner"? ... What makes an entertainer a "star"? ... When electricity is cut off, a region experiences a blackout—how might a blackout be compared to losing a loved one? ... How can grief be compared to one person maintaining the electricity of two? ... What does it mean to "keep connected," "keep a memory alive," "keep a light burning"? ... What does a lighthouse provide besides light? ...

***The City from Greenwich Village* [1922] by John Sloan**
Oil on canvas
National Gallery of Art, Washington, D.C.

Electric lights drape the darkened city like long strings of jewels, and the apartment buildings seem to levitate, as though upwards grounded. Yet, Sloan's vision of urban night is anything but bleak. Night in the city is like a second day. *Electricity* connects the disconnected, unifying individual people in their individual homes (maybe one of the reason blackouts are so terrifying is that we're afraid to be alone, cut off from access to information and from each other). Like other contemporary American painters belonging to the "Ashcan" school (William Glackens, George Luks, Robert Henri, Everett Shinn, and George Bellows), Sloan worked as an illustrator for newspapers and magazines—obviously, his real-life experiences at the city desk influenced the edgy, documentary-like charge of his work.

***The Autobiography of Benjamin Franklin* [1771-1788]**
by Benjamin Franklin

Franklin's experiments "for drawing lightning from the clouds"[5] were published in a pamphlet called "Experiments and Observations on *Electricity*." The work was universally admired for its ingeniousness. This delightful biography offers a colorful view of colonial life, and it is recommended for anyone who needs a lesson in the spirit of being a "true American." Franklin, one of 17 children, changed his life circumstances dramatically, rising from indentured printer to statesman to diplomat to inventor, as only an American could have done.

ENERGY 08

Like any other part of the environment, a belief environment can be fragile, composed of parts that are interconnected by both historical accidents and well-designed links. Consider, for instance, that delicate part of our belief environment concerned with the disposition of our bodies after death. Few people believe that the soul resides in the body after death—even people who believe in souls don't believe that. And yet few if any of us would tolerate a "reform" that encouraged people to dispose of their dead kin by putting them in plastic bags in the trash, or otherwise unceremoniously discarding them. Why not? Not because we believe that corpses can actually suffer some indignity. A corpse can no more suffer an indignity than a log can. And yet, the idea is shocking, repulsive. Why? The reasons are complex, but we can distill a few simple points for now. A person is not just a body; a person has *a body. That corpse is the body of dear old Jones, a Center of Narrative Gravity that owes its reality as much to our collaborative efforts of mutual heterophenomenological interpretation as to the body that is now lifeless. The boundaries of Jones are not identical to the boundaries of Jones's body, and the interests of Jones, thanks to the curious human practice of self-spinning, can extend beyond the basic biological interests that spawned the practice. We treat his corpse with respect because it is important for the preservation of the belief environment in which we all live.*

CONSCIOUSNESS EXPLAINED
DANIEL C. DENNETT

From: **EN-ER-GY** *n* (1599)
1 A : DYNAMIC QUALITY <NARRATIVE ~> B : THE CAPACITY OF ACTING OR BEING ACTIVE <INTELLECTUAL ~> 2 : VIGOROUS EXERTION OF POWER : EFFORT <INVESTING TIME AND ~> 3 : THE CAPACITY FOR DOING WORK 4 : USABLE POWER (AS HEAT OR ELECTRICITY); *ALSO*: THE RESOURCES FOR PRODUCING SUCH POWER

ENERGY

Action requires *energy*, different kinds of actions require different kinds of energy, and whenever an action occurs, one kind of energy transforms into another. In physics, energy is usually defined as the *capacity to do work*.

A basic distinction can be made between two forms of energy—*kinetic* and *potential*. Kinetic energy is the energy of *matter in motion*. A moving billiard ball can hit a ball at rest, thus setting it in motion. In the collision, the first ball exerts an almost instantaneous force on the second, causing it to accelerate for a very short time. The fact that the second ball moves tells us that the force has done work on it. A moving hammer also has kinetic energy, and it does work on a nail as the nail penetrates a wall. The amount of kinetic energy an object has depends upon its mass and its velocity. A light object traveling very quickly could have the same kinetic energy as a heavier object traveling less quickly. We can do the same work on a nail by striking it slowly with a very heavy hammer or faster with a lighter one.

Potential energy is another simple form of energy. When we pull back on a pendulum, we store potential energy in the pendulum by giving it the *capacity to fall under the influence of the force of gravity*. The amount of gravitational potential energy a body possesses depends on its height alone (its vertical position). As we lift it, we do work on it, and the amount of work we do is equal to the final potential energy of the pendulum. When the pendulum is left free to fall, it accelerates downward: it increases its speed. The speed it can reach at the bottom of the arc depends on the height from which it started. While the pendulum swings down, it loses potential energy as it lowers its position, but it gains kinetic energy as it increases its speed. *The kinetic energy it has at the bottom is equal to the decrease in potential energy from its maximum height to the lower position*. Energy is transformed from potential to kinetic. As the pendulum goes up again, its kinetic energy turns back into potential energy. If no force other than the force of gravity acts on the pendulum, its total energy—the sum of its kinetic energy and its potential energy—is always the same, in every position, at any time. It is *conserved*.

***See:* Field, Force, Friction, Gravity, Heat, Motion, Work**

$$E_K = \frac{1}{2} m v^2$$

[Eq. 3] ***The kinetic energy (E_K) of a body of mass m moving at speed v is equal to one half of the product of the mass times the square of the speed.***

THINK ABOUT IT: *Conservation of Energy*

The total energy of an isolated system is *conserved*—it does not change in time. A system is isolated when no external force is acting on it. In the case of the pendulum, the isolated system is given by the pendulum itself plus the Earth, which is responsible for the *gravitational force*. The total energy of the system is the sum of the kinetic energy of the pendulum plus its gravitational potential energy, which depends only on its height. If the pendulum slows down and then stops, it means that the system is not isolated. The slowing down of the pendulum happens because of air resistance. Including the air surrounding the pendulum as part of "the system" makes the system an isolated one because the energy lost by the pendulum is transferred to the air molecules. The total energy of the system given by Earth, pendulum, and air is conserved.

READ ABOUT IT: *Nuclear Hazards & Meaning in Physics*

Nuclear Madness: What You Can Do [1978] by Helen Caldicott — This important intellectual sounds out on the nuclear power industry, hazards to humanity, and the obligations of reasonable people to become informed about opaque government and corporate policies. Caldicott, an Australian pediatrician, formed Physicians for Social Responsibility (Nobel Prize, 1985), a coalition of health professionals who regard nuclear power as our greatest threat to safety. This classic *cri de coeur* is reminiscent of Rachel Carson's *Silent Spring*.

Physics for Poets [1970] by Robert March — This thoughtfully written resource is suitable for both scientist and nonscientist. According to March, a two-time winner of the American Institute of Physics Science Writing Prize, physicists and poets are engaged equally in a search for meaning. March leads the reader easily and logically from classical mechanics through relativity and quantum mechanics.

TALK ABOUT IT: *Beyond Physics—Energy & Ideals*

The energy of concentrates (medicines, perfumes, poisons) versus the energy of dilutions ... Idealism as a concentrated form of energy ... Is the energy of idealism kinetic or potential? ... Since idealism seeks to transcend the conditions of reality, it depends on its own absence. Like potential energy, it exists by virtue of its capacity *to cause change, by virtue of its "height alone" ... Art resolves the irony inherent to idealism by representing in time and space an ideal that might otherwise exist only in the mind of the idealist ... Idealism in the form of art can also be widely disseminated and understood ...*

The Kiss [1907-1908] by Gustav Klimt
Oil on canvas
Österreichische Galerie Belvedere, Vienna

The couple of this famous painting is wrapped in a cape that protects them as much as it enlivens those who stand beyond it. The exhibitionistic *energy* of the mosaic barrier originates in the suggestion of opposing positions—outside/inside, viewer/subject, conservative/modern. Klimt and his contemporaries ("The Vienna Secession") objected to moralistic conventions in Austrian art and culture. In *The First Moderns*, author William Everdell explains that *Modernismus*, the artistic trend to "present life as it was, low life as well as high, sexual as well as romantic,"[6] was being discredited and even banned at the time. Klimt, whose work was influenced by Japanese woodblock prints, forgoes depth of field in *The Kiss* to present a flattened, tree-like image that is as rooted and stable as it is mystifying and erotic, indicating perhaps that the supposed decadence of romantic love (and Klimt's imagery) is bound in the wholesome and the organic.

Consciousness Explained [1991] by Daniel C. Dennett

What is the nature of consciousness? Dennett deftly tackles this question, drawing from philosophy and biology to study perception in rich detail. The book contains anecdotes and exercises (e.g., "Imagine a purple cow."). More serious passages pertaining to where the mind is "contained," what constitutes conscious life, what the nature is of time and experience, and what happens to consciousness after death touch naturally upon the concept of *energy*.

ENTROPY 09

As nothing human is eternal, but every sublunary
object, especially the life of man, is always declining
from its origin to its decay; and Don Quixote had
no particular privilege from heaven, exempting
him from the common fate, the end and period of his
existence arrived, when he least expected its approach.

DON QUIXOTE
MIGUEL DE CERVANTES
TRANSLATED TOBIAS SMOLLETT

I met a traveller from an antique land,
Who said—"two vast and trunkless legs of stone
Stand in the desert ... near them, on the sand,
Half sunk a shattered visage lies, whose frown,
And wrinkled lips, and sneer of cold command,
Tell that its sculptor well those passions read
Which yet survive, stamped on these lifeless things,
The hand that mocked them, and the heart that fed;
And on the pedestal these words appear:
My name is Ozymandias, King of Kings,
Look on my Works ye Mighty, and despair!
Nothing beside remains. Round the decay
Of the colossal Wreck, boundless and bare
The lone and level sands stretch far away."

OZYMANDIAS
PERCY BYSSHE SHELLEY

From: **EN-TRO-PY** *n* (1875)
1 : A MEASURE OF THE UNAVAILABLE ENERGY IN A CLOSED THERMODYNAMIC SYSTEM THAT IS ALSO USUALLY CONSIDERED TO BE A MEASURE OF THE SYSTEM'S DISORDER AND THAT IS A PROPERTY OF THE SYSTEM'S STATE AND IS RELATED TO IT IN SUCH A MANNER THAT A REVERSIBLE CHANGE IN HEAT IN THE SYSTEM PRODUCES A CHANGE IN THE MEASURE WHICH VARIES DIRECTLY WITH THE HEAT CHANGE AND INVERSELY WITH THE ABSOLUTE TEMPERATURE AT WHICH THE CHANGE TAKES PLACE; *BROADLY* : THE DEGREE OF DISORDER OR UNCERTAINTY IN A SYSTEM 2 A : THE DEGRADATION OF THE MATTER AND ENERGY IN THE UNIVERSE TO AN ULTIMATE STATE OF INERT UNIFORMITY B : A PROCESS OF DEGRADATION OR RUNNING DOWN OR A TREND TO DISORDER 3 : CHAOS, DISORGANIZATION, RANDOMNESS

ENTROPY

Entropy is said to be the measure of disorder in a system. Consider, for instance, the randomly-moving air molecules in a two-room apartment. The natural tendency, of course, is for the molecules to disperse evenly between the rooms. If we were to force all the molecules into one room, this arrangement would correspond to a *higher order and a lower entropy* than an even distribution would because at least we would know in which room any single molecule is. In principle, even with the door open, all the molecules could go back into just one room, by virtue of their random motion. However, such an event would have such a low probability that it is practically impossible. In practice, air molecules will "choose" the state that corresponds to the highest disorder.

Entropy can also be seen as a measure of how much is unknown about a system. Imagine we are throwing three normal dice, but for some reason, we cannot see how each die falls. We only know the sum of their points. If one sum is 3, there is only one way the dice can be—all 1s—so we know everything. If a total is 4, there are three ways the dice could be, two 1s and a 2, though we don't know which die has the 2. If a total is 10, there are 27 different ways the dice could be. This number, 27, measures the unknown. It describes how many different throws of the dice could have resulted in a total of 10. The real state of the system, in this case, the face-up number on each die, is called the *microstate*, and what we actually know about the system, in this case, the total, is called the *macrostate*. Here, entropy is the number of microstates compatible with a given macrostate.

And why do we only know the total? Of course, if we saw the dice fall, we would know the number on each side, but real-world systems are generally composed of too many components to know more than a few large-scale "totals." A cup of water, for instance, has too many molecules (on the order of millions of billions of billions) for anyone to know where each is, where it is going, and how fast. However, it is relatively easy to determine the volume and temperature. Here, entropy is a measure of how many ways the water molecules can be arranged while retaining that same volume and temperature.

***See:* Chaos, Energy, Heat, Particle**

$$S = k \ln \Omega$$

[Eq. 4] ***The entropy of a system (S) is given by the product of the Boltzmann constant (k) times the natural logarithm of the number (Ω) of microstates compatible with a given macrostate.***

THINK ABOUT IT: *Entropy & Work*

In *thermodynamics*, it is possible to turn mechanical energy into heat completely, but it is impossible to turn heat completely into mechanical energy. When we slide a book across a table, the force of friction "does work" on the book by turning its initial kinetic energy into heat. The sliding book is an "ordered" form of motion, whereas the final increase in temperature of the molecules of the book and of the table is a measure of a "disordered" form of motion. During the process, the entropy of the whole system increased, but the total energy stayed the same. Thus, entropy can be defined as the measure of the energy of a system that is unavailable to do work.

READ ABOUT IT: *Structure, Sequence, The Second Law of Thermodynamics, & The Feynman Lectures*

Does God Play Dice?: The New Mathematics of Chaos [1989] by Ian Stewart — Stewart, a Michael Faraday Award-winner offers a compelling study of the history and nature of chaos. The book refers to such things as heredity, gambling, statistics, and, of course, entropy. In an anecdote about his children, Stewart explains that the two characteristics of chaos are *mixing* and *expansion*.

Entropy and Art: An Essay on Disorder and Order [1971] by Rudolf Arnheim — Arnheim, a 20th-century theorist, discusses entropy as a measure of structure versus a measure of sequence. He writes carefully and elegantly about the second law of thermodynamics in relation to the human need for order, drawing from physics, philosophy, and psychology (e.g., Why *do* people strive for order?).

The Feynman Lectures on Physics [1963] by Richard Feynman, Robert Leighton, and Matthew Sands — This enormous three-book set covers everything from entropy to Newtonian mechanics to quantum physics. The series is expensive but worth the investment for the serious student—otherwise, ask for it at your library.

TALK ABOUT IT: *Beyond Physics—Entropy & Time*

Loss of heat, order, structure ... What is the difference between the deterioration of manufactured forms versus that of natural forms? ... Is decay always undesirable? ... Can entropy be seen as the progression from whole to particulate? ... Are rare objects precious because they cannot be remade or regenerated? ... Can extinction be described in relation to entropy? ... The greenhouse effect? ... Does entropy indicate a downward and forward movement of time? ...

***The Last Supper* [1495-1498] by Leonardo da Vinci**
Tempera mural
Santa Maria della Grazie, Milan

The guarantee of impermanence is easier to accept with objectivity in science than in art. Da Vinci's masterpiece began to deteriorate soon after completion because the artist experimented with an oil-tempera medium that did not adhere well to the wall. Efforts in the centuries since to preserve the fresco by over-painting and/or gluing fallen flakes have often caused more damage than good. Now, experts forestall degradation (*entropy*) by maintaining "atmospheric" equilibrium: small visitor groups are admitted for brief periods. What makes this work so special? It was Da Vinci's study of physics and mechanics that affected the unique quality of his vision.

***Don Quixote* [1605] by Miguel de Cervantes**
Translated by Tobias Smollett

Don Quixote's death—the end of his journey. Surrounded by friends, he accepts his fate with grace. His life was well-lived, rich with optimism, adventure, and soul-searching. Is dying really an end when one leaves the world positively transformed? Can't death (like *entropy*) be considered as an essential part of nature's arrangement?

"Ozymandias" [1818] by Percy Bysshe Shelley

This unsettling poem about the statue of the forgotten king of a lost empire is evocative of the effects of *entropy*. It reminds us that human works will *not* last forever. Shelley was married to Mary Shelley, author of *Frankenstein* and daughter of Mary Wollstonecraft, who was herself the author of *A Vindication of the Rights of Woman* (See: *Position*), one of the first books to address issues of feminism.

EQUATION 10

What is really peculiar to quantities is that we compare or contrast them in terms or on grounds of equality. We predicate 'equal,' 'unequal,' of all of the quantities mentioned. One solid is equal to another, another, per contra, *unequal. We use these terms also of time in comparing the periods of it. So also of all other quantities that we have previously mentioned. Of nothing, moreover, save quantities can we affirm these two terms. For we never say this disposition is 'equal' to that or 'unequal.'*

We say it is 'like' or 'unlike.' One quality —whiteness, for instance—is never compared with another in terms or on grounds of equality. Such things are termed 'like' and 'unlike.' Thus our calling something 'equal,' 'unequal,' is the mark, above all marks, of quantity.

THE CATEGORIES: ON INTERPRETATION: PRIOR ANALYTICS
ARISTOTLE
TRANSLATED BY H. P. COOKE

From: **EQUA·TION** *n* (14C)
1 A : THE ACT OR PROCESS OF EQUATING B (1) : AN ELEMENT AFFECTING A PROCESS : FACTOR (2) : A COMPLEX OF VARIABLE FACTORS C : A STATE OF BEING EQUATED; *SPECIF* : A STATE OF CLOSE ASSOCIATION OR IDENTIFICATION 2 A : A USUALLY FORMAL STATEMENT OF THE EQUALITY OR EQUIVALENCE OF MATHEMATICAL OR LOGICAL EXPRESSIONS B : AN EXPRESSION REPRESENTING A CHEMICAL REACTION QUANTITATIVELY BY MEANS OF CHEMICAL SYMBOLS

$E\left(\frac{1}{n-1}\sum_{i=1}^{n}\left(y_i-b\overline{x_i}\right)^2\right)$
$=\sum_{i=1}^{n}\sigma^2\frac{\left(S-x_i^2\right)}{(n-1)S}$
$=\frac{\sigma^2}{(n-1)S}\left(nS-\sum_{i=1}^{n}x_i^2\right)$
$=\sigma^2$

EQUATION

ALL PHYSICAL LAWS CAN BE expressed in words, but throughout the history of physics, scientists have found increasingly compact ways to express them by means of *equations*. Equations allow for a fundamental economy of thought. Whereas it requires an entire book to explain *Albert Einstein's* [1879-1955] equations for general relativity, the theory itself is contained in just one equation. If we are familiar with the language of mathematics, we can immediately recognize the meaning and main features of a physical theory from its equations. Every physical quantity and almost every physical law can be defined and expressed in mathematical terms. If an equation cannot be found, there is the impression that the understanding of a given physical phenomenon is not complete. Also, the experiments that test the validity of physical theories involve numbers (values of measured and predicted quantities) and equations to find them. Science as we know it began with the mathematical approach to the study of nature.

Common sense suggests that the more a spring is compressed, the more effort is needed to keep it compressed, but common sense does not indicate *how much* effort, nor does it provide an *objective way to quantify* such effort. To say that the *elastic force exerted by a compressed spring increases as we compress it*, is to make a *qualitative* statement. But to write $F = k\,x$, where F is the magnitude of the force, x is the length by which the spring is compressed, and k is a constant that depends on the particular spring, is to make a very precise mathematical statement: *the force is proportional to the compression*. If x doubles, the force doubles—*exactly*.

Other equations in physics have a more fundamental nature. The equation $F = m\,a$ does not describe a particular force; rather, it gives a well-defined *relation between force and acceleration*. This equation does not simply say that a force applied to a body is related in some way to its motion, but it indicates that an applied force results in an acceleration. At the same time, the equation provides a way to compute such acceleration, given the values of the force and of the mass of the body. Solving an equation means finding the value of one of its quantities once you know the value of each of the others.

See:* *Measurement

THINK ABOUT IT: ***Physics & Mathematics***

For the most part, learning physics entails learning how to make use of a few basic equations and how to apply them to particular situations. One reason physicists enjoy their discipline is that a relatively small number of fundamental concepts and related equations have to be kept in mind. The majority of the equations that appear in physics books are "derived" from fundamental ones. For example, the motion of projectiles, pendulums, falling apples, and orbiting planets can be studied by means of two equations: the *second law of dynamics* and the *law of gravity*.

READ ABOUT IT: ***Number Games, The Life of Einstein, & An Illustrated Guide to Relativity***

An Adventurer's Guide to Number Theory [1968] by Richard Friedberg — In this classic book of number theory, readers are encouraged to maintain a playful outlook toward math. Friedberg describes the number 10 as "smooth like a pebble on the beach."

Albert Einstein: A Biography [1993] by Albrecht Fölsing — Fölsing's biography of this genius and humanitarian has hundreds of references to primary source material. The details of Einstein's professional and personal life are woven into the tumultuous events of the 20th century. Direct accounts of his meetings with royalty, dignitaries, and fellow scientists make for fascinating reading.

Relativity Simply Explained [1962] by Martin Gardner — This book explains Einstein's theories (and *equations*) for general readers. The clarity of the writing (it has been updated several times) and Anthony Ravielli's 1950s-style graphic illustrations make this a classic.

TALK ABOUT IT: ***Beyond Physics—Equations & Human Experience***

Equations require quantifiable variables ... What sorts of things in life are unquantifiable? ... In regard to people, how can circumstances and objects be equated with states of being (e.g., material goods with happiness, marriage with fulfillment, jobs with security, etc.) when individual human experience is so unique? ... Television stories are "quantified" in order to meet predetermined time limits ... Such limits are used to guarantee sustained audience interest that might lead to ad sales ... How do independent forms of music, film, photography, and dance satisfy a need for "unquantifiable" stories? ... How are equations similar to logos? ... Comic strips? ...

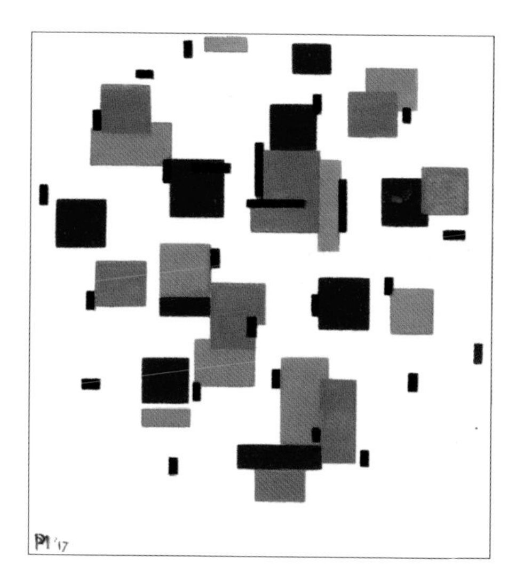

***Composition in Color B, 1917* [1917] by Piet Mondrian**
Oil on canvas
Kröller-Müller Museum, Otterloo, The Netherlands

Mondrian's subject matter evolved from the landscapes and still lifes typical of Dutch Expressionism and Symbolism to pure abstractions characteristic of the Cubist and De Stijl movements. The patchwork paintings for which he is best remembered are like visual *equations*. Despite their apparent simplicity, the theory behind these color grids is carefully wrought: the artist felt that the removal of any one component would cause the whole to collapse. According to Janson's *History of Art*, the artist removed all "objects" belonging to nature or humanity from his work, hoping to achieve "equilibrium through opposition."[7] There is something promising about the harmonious severity of Mondrian's compositions: like a snapshot of city streets taken from a bird's-eye view or like a child's stop-motion game, they suggest playful new applications for consciousness with the advantage of a fresh, frozen-in-time perspective.

The Categories: On Interpretations: Prior Analytics
[350 B.C.E.] by Aristotle, Translated by H. P. Cooke

The Greek philosopher and student of Plato lays out his criteria for interpretation in this classic text. Despite Aristotle's formidable attempt to express natural phenomena with a degree of objectivity, it is impossible to do so without the aid of *equations*. Still, this is an important and inspiring book to read—the discipline of physics arose from that of natural philosophy, and the ideas Aristotle expresses within it have the mark of originality, innovation, and intelligence.

FIELD

I ran down the hill in that vacuum of crickets like a
breath travelling across a mirror she was lying in the
water her head on a sand spit the water flowing about
her hips there was a little more light in the water her
skirt half saturated flopped along her flanks to the
waters motion in heavy ripples going nowhere renewed
themselves of their own movement I stood on the bank I
could smell the honeysuckle on the water gap the air
seemed to drizzle with honeysuckle and with the rasping of
crickets a substance you could feel on the flesh
is Benjy still crying
I don't know yes I don't know
poor Benjy
I sat down on the bank the grass was damp a little
then I found my shoes wet
get out of that water are you crazy
but she didn't move her face was a white blur framed
out of the blur of the sand by her hair
get out now
she sat up then she rose her skirt flopped against
her draining she climbed the bank her clothes flopping
sat down
why don't you wring it out do you want to catch
cold
yes
the water sucked and gurgled across the sand spit and
on in the dark among the willows across the shallow the
water rippled like a piece of cloth holding still a little
light as water does.

THE SOUND AND THE FURY
WILLIAM FAULKNER

From: **FIELD** *n* (BEFORE 12C)
1 D : A LARGE UNBROKEN EXPANSE (AS OF ICE) 2 A : AN AREA OR DIVISION OF AN ACTIVITY B : THE SPHERE OF PRACTICAL OPERATION OUTSIDE A BASE (AS A LABORATORY, OFFICE, OR FACTORY) 6 A : A REGION OR SPACE IN WHICH A GIVEN EFFECT (AS MAGNETISM) EXISTS.

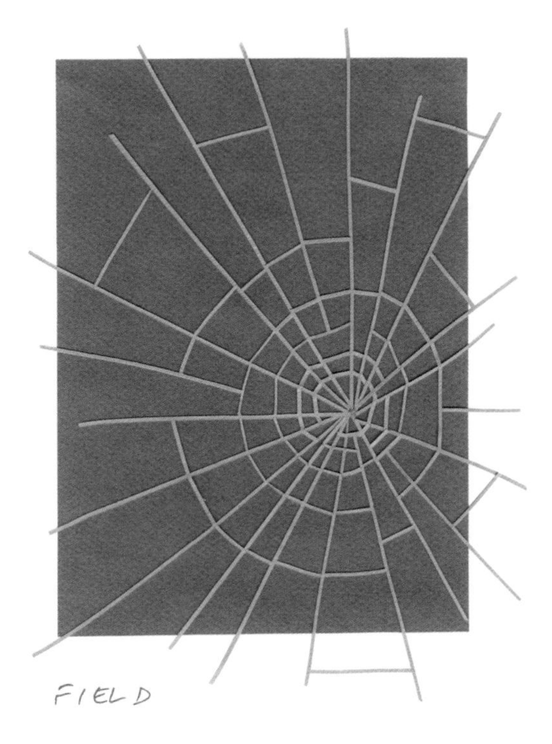
FIELD

THE WORD *FIELD* SUGGESTS SOMETHING that fills space, something that is everywhere. The concept was introduced in physics at the beginning of the 19th century by *Michael Faraday* [1791-1867] to describe how electric charges interact. What was then known is that two electric charges attract or repel each other with a force that depends on the value of their charges and on the distance between them. What was unclear is the way in which one charge detects the presence of the other—in particular, when their distance changes and when there is nothing but empty space between them.

Faraday assumed that an electric charge creates an electric field in every point of the space surrounding it; a field more intense when closer to the charge and weaker when farther away. In the presence of an electric charge, the space is not empty; it is filled with an electric field, and ultimately, with energy. Consider two electrically charged particles, A and B. Physicists now think about their interaction in terms of particle B "feeling" the electric field created by particle A and then being attracted toward A by a force proportional to its charge and to the intensity of the field in the exact position where particle B is located. Particle B knows nothing about the presence of particle A; it just interacts with the field in its own location. The same holds true for particle A: it "feels" the field created by particle B.

If particle A moves away from particle B, the force between them changes, since it is dependent upon their distance. Does B "feel" the change in the force immediately? As A starts to move, its electric field changes, starting from the position of A. This small change, or *perturbation*, in the electric field *propagates* in space, eventually reaching B and causing it to experience a change in the electric force after a certain time. Such propagation is similar to waves on the surface of a lake that move out to the edge and then reach the shore. Moreover, if A *oscillates*, its electric field oscillates, and an oscillating electric field induces an oscillating magnetic field, which, in turn, induces an oscillating electric field, and so on. This propagating oscillation is called an *electromagnetic wave*. Light is a propagating oscillation of the electromagnetic field at frequencies that our eyes can perceive.

***See:* Electricity, Force, Gravity**

THINK ABOUT IT: ***Field Theories***

The concept of *field* has many applications in physics. A room's temperature can be described as a field—for example, in winter, the temperature will have a larger value in positions close to the heater and a lower value nearer to an open window. The velocity of wind can be thought of as a *vector field*: it is represented on weather maps by arrows signifying direction and intensity. Interactions between particles are also described in terms of fields, and, in fundamental physics, elementary particles themselves are described by fields in what are called quantum field theories.

READ ABOUT IT: ***Thinking in Pictures, Networks, & Connections in Living Systems***

The Emperor's New Mind [1989] by Roger Penrose — Penrose, a mathematical physicist from Oxford University, examines the mind/body divide in relation to artificial intelligence. In "Can a Computer Have a Mind?" and "Algorithms and Turning Machines," he discusses matter-transmission and thinking in pictures.

Linked: How Everything Is Connected to Everything Else [2003] by Albert-László Barabási — As humanity moves into unprecedented stages of connectedness, it faces new possibilities and new dangers. Barabási talks about *networks*—from the Internet to corporations to terrorists—to show how these are linked according to natural laws.

The Web of Life: A New Scientific Understanding of Living Systems [1996] by Fritjof Capra — Capra is the insightful author of nonfiction bestsellers *The Tao of Physics* and *The Turning Point*. *The Web of Life* succeeds in making important connections among science, written language, math, biology, chemistry, ecology, and the mind.

TALK ABOUT IT: ***Beyond Physics—Field & Intuition***

Field theory describes interactions between bodies across space ... Can human interactions also "move" across fields? ... Prayer? ... Intuition? ... Love? ... What is the "medium" through which such exchanges might travel? ... What does the concept of an unbroken expanse—that is, a field—imply about time; e.g., how might Object A's "recent past" becomes Object B's "about to happen"? ... In fiction, stream-of-consciousness writing is a narrative device that indicates connection rather than difference or separation ... What might an author gain by using the stream-of-consciousness style? ... What is the impact on the reader's understanding? ...

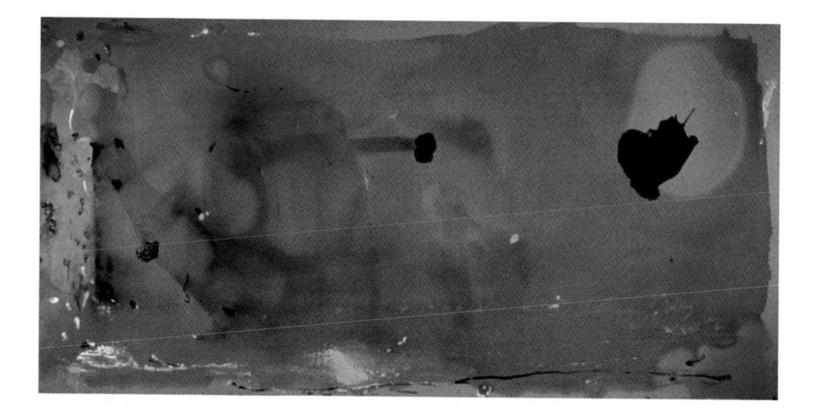

***Madame Matisse* [1983] by Helen Frankenthaler**
Acrylic on canvas
Private Collection

Frankenthaler, an American Abstract Expressionist painter, was part of the "Color Field" movement that began in the 1940s. She stained unprimed canvases with thin color washes to create forms whose shape and tonal density depended on the absorbency of the surface in use. There is a fluid, relaxed quality to *Madame Matisse* that leads to a viewing experience more personal than objective. The eye moves through a *field* of color in which an event in one location could conceivably lead to events in others. Frankenthaler was married to artist Robert Motherwell (See: *Velocity*).

***The Sound and the Fury* [1929] by William Faulkner**

The *Sound and the Fury* is a gorgeous *field* of words, woven through a narrative that itself is nonlinear—when one sentence appears, the reader gets the sense that others are affected, magically, miles away. Even Faulkner's unusual layout contributes to the feeling of fluidity:

I stood on the bank I
could smell the honeysuckle on the water gap the air
seemed to drizzle with honeysuckle and with the rasping of
crickets a substance you could feel on the flesh
is Benjy still crying
I don't know yes I don't know
poor Benjy—

Faulkner, a Southerner who wrote primarily of the demise of the American South, finished this novel when he was just 22 years old. It is amazing that someone so young could have produced a work of such mature depth. Faulkner's artistic bravery and secure knowledge of place allowed him to conduct the bold experiments with language and style for which he is most widely remembered.

FORCE 12

I write. My mother was a Florentine,
Whose rare blue eyes were shut from seeing me
When scarcely I was four years old, my life
A poor spark snatched up from a failing lamp
Which went out therefore. She was weak and frail;
She could not bear the joy of giving life,
The mother's rapture slew her. If her kiss
Had left a longer weight upon my lips
It might have steadied the uneasy breath,
And reconciled and fraternised my soul
With the new order. As it was, indeed,
I felt a mother-want about the world,
And still went seeking, like a bleating lamb
Left out at night in shutting up the fold,—
As restless as a nest-deserted bird
Grown chill through something being away, though what
It knows not. I, Aurora Leigh, was born
To make my father sadder, and myself
Not overjoyous, truly.

AURORA LEIGH
ELIZABETH BARRETT BROWNING

From: **FORCE** *n* (14C)
4 A : AN AGENCY OR INFLUENCE THAT IF APPLIED TO A FREE BODY RESULTS CHIEFLY IN AN ACCELERATION OF THE BODY AND SOMETIMES IN ELASTIC DEFORMATION AND OTHER EFFECTS B : ANY OF THE NATURAL INFLUENCES (AS ELECTROMAGNETISM, GRAVITY, THE STRONG FORCE, AND THE WEAK FORCE) THAT EXIST ESPECIALLY BETWEEN PARTICLES AND DETERMINE THE STRUCTURE OF THE UNIVERSE

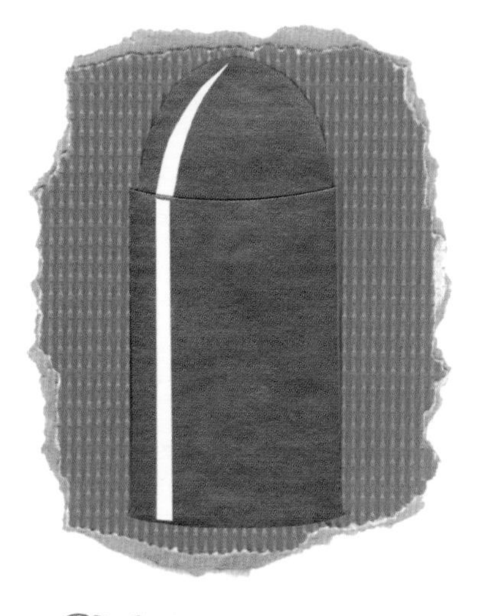

FORCE

WHENEVER AN OBJECT IS SET in motion—pushed, shaken, twisted, or stopped—it is experiencing a *force*. More precisely, forces cause objects to accelerate. If no force is acting on an object, or if the sum of all the forces acting on it is zero, then the object will keep its velocity constant in speed and direction. This is called the *principle of inertia*. Since every motion on Earth eventually comes to an end, it is natural to think that something is needed to keep things moving, but a *force is not the cause of motion*: in the absence of external forces, every object has a natural tendency to keep moving, forever. A force is responsible only for a *change in velocity*.

In everyday life, it is difficult to observe the behavior of an object when a single force or no force is acting on it. Often, more than one force is present at the same time. Even the simple act of holding and then throwing a ball involves a number of forces. We make an effort simply by keeping the ball in our hands—*we apply a force opposite to the force of gravity* so that the total force on the ball is zero. The ball, initially at rest, stays at rest—that is, it has no acceleration. In order to describe the act of throwing a ball, we need to consider at least three forces acting at the same time: gravity, our "push," and air resistance. As the ball starts to move, it needs to make its way through the molecules of air around it. This action results in *air resistance*, a force that always opposes motion and that increases as the speed of the moving object increases. To throw the ball upward, the force we have to apply needs to be larger than the sum of the force of gravity *plus* the force due to air resistance. If this condition is met, the ball accelerates up. As it leaves our hand, only gravity and air resistance will act on it. The ball will then accelerate downward—namely, *it will move up, slowing down, and then move down with increasing speed*. Air resistance is still present, and it increases as the speed increases. If there is enough space for the ball to fall, it will eventually reach a speed where the force due to air resistance will be as intense as the force of gravity. From that moment, the total force on the ball is zero, and the ball will fall with constant velocity. When we want to use Newton's second law of dynamics, we must take into account all the forces acting on a given object.

See:* *Acceleration, Friction, Gravity, Mass, Momentum, Velocity

THINK ABOUT IT: *Newton's Laws of Dynamics*

From the English translation of Newton's original in Latin:
I. Every body perseveres in its state of being at rest, or of moving uniformly straightforward except insofar as it is compelled to change its state by forces impressed. [*The principle of inertia. The velocity of a body will stay constant in magnitude and direction unless a force acts on it.*]
II. A change in motion is proportional to the motive force impressed and takes place along the straight line in which that force is impressed. [*The acceleration experienced by a body is proportional to the total force acting on it, and it has the same direction as the force.*]
III. To any action there is always an opposite and equal reaction; in other words, the actions of two bodies upon each other are always equal and always opposite in direction. [*When we throw a ball, we apply a force to it, while the ball applies a force to our hand. It would take less effort to move an empty hand.*]

READ ABOUT IT: *Newton, The Principia, & Archimedes*

Newton: Texts, Backgrounds, Commentaries [1995] edited by I. Bernard Cohen and Richard Westfall — In this account of Newton's career, Newton's original writings are accompanied by commentary in order to offer some historical perspective on his accomplishments.

The Principia: Mathematical Principles of Natural Philosophy [1687] by Sir Isaac Newton — Newton lays out in mathematical terms the principles of time, force, and motion in this incredible classic text. The ideas show the workings of an ambitious, creative mind. Though the book is complicated and perhaps too expensive for any but the most serious student, it is available to the curious at libraries.

The Works of Archimedes [1897] by T. L. Heath — Heath's translation of the works of one of history's greatest engineers and inventors demonstrates that many of the initial pursuits of physics remain in place despite remarkable advances in scientific knowledge.

TALK ABOUT IT: *Beyond Physics—Force & Altruism*

Consider some of the forces that motivate people—love, grief, duty, desire, ambition, loyalty, guilt, fear of death ... What might move someone toward altruism? ... What inspires an individual to give his or her life to something larger? ... Is self-sacrifice always an act of courage? ... Why would someone want to be considered a hero, even past the point of death? ... Altruism versus compassion? ...

The Tragedy [1903] by Pablo Picasso
Oil on wood
National Gallery of Art, Washington, D.C.

Following the suicide of his friend Carles Casagemas, Picasso entered what has come to be known as his "Blue Period." "Blue" refers to the dominant color used in the works and to the pathos of the artist's imagery. Whatever the origin of his dark emotions, we are the lucky heirs of a grief that produced some of the greatest—and riskiest—artwork of the century. This masterpiece depicting three ravaged figures in front of the sea at Barcelona demonstrates the *force* of tragedy and the *force* of Picasso's singular creative genius—he painted *The Tragedy* at a mere 22 years of age. The "Blue Period" was followed by artist's "Rose Period;" an indication, perhaps, of better times.

Aurora Leigh [1856] by Elizabeth Barrett Browning

What a beautiful line: *The mother's rapture slew her*. Though Aurora Leigh's mother could not bear the *force of life*, she had the *force of effect*—her death set into motion the artistic destiny of Aurora, the heroine. Author Virginia Woolf called this book "a masterpiece in embryo."[8] And truly, Browning's courageous and touching work endures as a detailed testament to creative life, to female life, and to social and political life in the 1850s. It is said to have been Susan B. Anthony's favorite book, and it was much loved by Rudyard Kipling, Oscar Wilde, Christina Rossetti, and even Queen Victoria.

FRICTION 13

It would be impossible for any craftsman or sculptor
no matter how brilliant ever to surpass the grace
or design of this work or try to cut and polish the marble
with the skill that Michelangelo displayed. For the Pietà was
a revelation of all the potentialities and force of the art
of sculpture. Among the many beautiful features (including
the inspired draperies) this is notably demonstrated
by the body of Christ itself. It would be impossible to find
a body showing greater mastery of art and possessing
more beautiful members, or a nude with more detail
in the muscles, veins, and nerves, stretched over their
framework of bones, or a more deathly corpse.

The lovely expression of the head, the harmony
of the joints and attachments of the arms, legs,
and trunk, and the fine tracery of pulses and veins
are all so wonderful that it staggers belief that the
hand of an artist could have executed this inspired and
admirable work so perfectly and in so short a time.
It is certainly a miracle that a formless block
of stone could ever have been reduced to a perfection
that nature is scarcely able to create in the flesh.

LIVES OF THE ARTISTS: VOLUME I
GIORGIO VASARI
TRANSLATED BY GEORGE BULL

From: **FRIC-TION** *n* (1704)
1 A : THE RUBBING OF ONE BODY AGAINST ANOTHER B : THE FORCE THAT RESISTS RELATIVE MOTION BETWEEN TWO BODIES IN CONTACT 2 : THE CLASHING BETWEEN TWO PERSONS OR PARTIES OF OPPOSED VIEWS : DISAGREEMENT

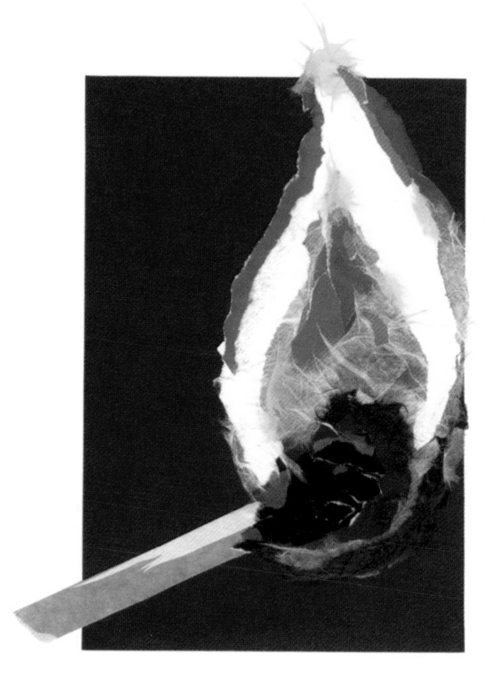

FRICTION

IMAGINE TRYING TO MOVE A heavy block of wood across the floor without lifting it. If the force we apply on the block is too small the block will not move—it will experience no acceleration. According to the second law of dynamics, if the acceleration of an object is zero, the total force applied on it must be zero as well. We can deduce that our force is not the only force acting on the block and that there is another force acting on it that is *equal and opposite in direction* to the one we are applying. Such a force is called the *force of static friction*. Static friction keeps something in place on a surface when the object is not pushed hard enough. The force of static friction is always equal to the force we apply, but it has a maximum value. If we apply a force larger than such a maximum, we can overcome static friction, and the block will move. In general, this maximum is larger for heavier objects, but it also depends on the roughness of the touching surfaces—in the case of this example, those surfaces would be the block and the floor.

Once the block starts to move, we need to apply a force to keep it in motion. If the block moves along a straight line at constant speed, then its acceleration is zero. Its speed is not increasing or decreasing, nor is its trajectory changing direction. Again, if the acceleration is zero, then the total force on the block must be zero. Since we are definitely applying a force, there must be another force that is equal in intensity but that acts in the opposite direction to cancel ours out. Such a force is called *force of kinetic friction*—the friction acting on objects in motion. It is important to realize that the particular motion of the block does not depend only on the force we apply, but on all the forces applied, including friction. If we could kick the block, it would slide along the floor until it stopped. In this particular case, kinetic friction would be the only force acting on the block, and it would always oppose the block's motion. The block *decelerates*; that is to say, it experiences acceleration in the direction opposite to the direction of motion. On the other hand, if there were no kinetic friction, the block could move across the room at constant velocity without our pushing it. The block would keep constant its velocity in speed and direction, according to the principle of inertia.

***See:* Acceleration, Energy, Force, Mass, Momentum, Motion**

THINK ABOUT IT: *Friction & Work*

When a block of wood is kicked across a floor, it will slide until its speed decreases and eventually goes to zero—and the block stops. The block started with some *initial kinetic energy* that decreased as the block's speed decreased. What happened to its initial energy? While the block was sliding, the only force acting on it—along the *direction of motion*, but opposite to it—was the force of kinetic friction. This force "did work" on the block until the block stopped: the total amount of work done was equal to the block's initial kinetic energy. The initial kinetic energy of the block did not disappear; it turned into heat. In fact, the temperature of the two surfaces that were in contact—the block and the floor—increased. This is why sandpaper gets hot when used.

READ ABOUT IT: *The Collapse of Time, Rare Tools, & Experiments for Beginners*

ABC of Relativity [1925] by Bertrand Russell — Russell's gentle clarification of Einstein's theories is authoritative and indispensable. Anyone trying to grasp "the collapse of one all-encompassing time" will enjoy chapters such as "What Happens and What is Observed."

The Physics of Skiing [2003] by David Lind and Scott Sanders — This is a fun way to engage in scientific talk about snow, equipment, and technique—skiing is just the motion of a body down an inclined plane!

Tools Rare and Ingenious: Celebrating the World's Most Amazing Tools [2004] by Sandor Nagyszalanczy — This picture book examines the history, function, and design of tools (e.g., dividers, gauges, planes, calipers, etc.), and it celebrates the genius of humankind.

Why Toast Lands Jelly-Side Down: Zen and the Art of Physics Demonstration [1997] by Robert Ehrlich — This book of user-friendly experiments is a solid learning tool for beginners, with illustrations of concepts ranging from dynamics to magnetism to optics to friction.

TALK ABOUT IT: *Beyond Physics—Friction & Dilemma*

Imagine a personal dilemma as a type of friction ... Consider the problem of choice ... Is it ever possible to know exactly "what to do"? ... What is free will, and how do we use or abuse it? ... Internally, we can appeal to our accumulated knowledge and our moral understanding of the world to help us select a path of action, but, externally, how can we separate sound advice from manipulation? ...

Vairumati [1789] by Paul Gauguin
Oil on canvas
Musée d'Orsay, Paris

In *Vairumati*, Gauguin establishes a striking evenness of gaze, the nature of which tends to escape the artistic products of most meetings between worlds. His celebration of the "primitive" splendor of Tahiti, where he spent the last years of his life, did not necessarily exploit the sensuality of Tahitian culture so much as invert the rigid constructs of beauty and propriety of his own. The *friction* of this painting lies more in the position of forced humility in which he places the Western observer than in a voyeuristic tension between artist and model. This practice of using the canvas as a "place" for the exchange of ideas was achieved in Gauguin's final work, *Where Do We Come From? What Are We? Where Are We Going?* [1897-1898]. Gauguin, who turned from banking to painting at the age of 35, received no formal training. Perhaps this is the reason he was able to cultivate a completely original style.

Lives of the Artists: Volume I [1550] by Giorgio Vasari
Translated by George Bull

The triumphs and struggles of the giants of the Renaissance (da Vinci, Brunelleschi, Botticelli, et al.) are described in detail by a knowledgeable contemporary. The vision of Michelangelo sanding the *Pietà* (See: *Mass*) to idealized perfection is tremendously moving; especially considering that the by-product of that *friction* was the heat of the hands of a genius. We are indebted to Vasari for recording an important time in the history of human ideas.

GRAVITY 14

Guilt, remember, is the bind that man experiences when he is humbled and stopped in ways that he does not understand, when he is overshadowed in his energies by the world. But the misfortune of man is that he can experience this guilt in two ways: as bafflement from without and from within—by being stopped in relation to his own potential development. Guilt results from unused life, from "the unlived in us."

THE DENIAL OF DEATH
ERNEST BECKER

From: **GRAV-I-TY** *n* (1509)
1 A : DIGNITY OR SOBRIETY OF BEARING B : IMPORTANCE, SIGNIFICANCE 2 : WEIGHT 3 A (1) : THE GRAVITATIONAL ATTRACTION OF THE MASS OF THE EARTH, THE MOON, OR A PLANET FOR BODIES AT OR NEAR ITS SURFACE (2) : A FUNDAMENTAL PHYSICAL FORCE THAT IS RESPONSIBLE FOR INTERACTIONS WHICH OCCUR BECAUSE OF MASS BETWEEN PARTICLES, BETWEEN AGGREGATIONS OF MATTER (AS STARS AND PLANETS), AND BETWEEN PARTICLES (AS PHOTONS) AND AGGREGATIONS OF MATTER, THAT IS 10^{39} TIMES WEAKER THAN THE STRONG FORCE, AND THAT EXTENDS OVER INFINITE DISTANCES BUT IS DOMINANT OVER MACROSCOPIC DISTANCES ESPECIALLY BETWEEN AGGREGATIONS OF MATTER—CALLED ALSO *GRAVITATION, GRAVITATIONAL FORCE*

GRAVITY

FOR MOST OF HUMAN HISTORY, the falling of an apple and the motion of the Moon in the sky were thought to be very different things, requiring different explanations. What Newton realized is that the force that makes an apple fall is the same force that keeps the Moon in its orbit. With the discovery of the *universal law of gravity*, Newton achieved one of the great goals of science: to understand and explain seemingly disconnected phenomena with a single simple formula. By saying that every mass attracts every other mass with a force given by his famous equation, Newton provided a theory that could describe how the apple falls and how the planets orbit, as well as how stars are bound in galaxies, why planets have a spherical shape, and the cause of the tides.

The law of gravity states that two masses are attracted by a force proportional to the masses themselves and inversely proportional to the distance between them. The force is more intense between larger masses. Also, the force between two bodies increases as they get closer to each other. This behavior is similar to the behavior of the force between two charges—the electric force is proportional to the product of the charges and is dependent upon their distance in the same way. But there are two important differences between the two forces. First, the gravitational force is always attractive—there are no positive and negative masses: *every mass always attracts other masses*. Second, the force of gravity depends upon the mass of the object! This is not trivial. *The mass in the gravitational force plays the role of the charge in the electric force*. But mass is also a measure of inertia, as expressed by the second law of dynamics. More massive objects have more inertia—this means it is harder to accelerate them. In the case of gravitational attraction, the mass provides a measure of the inertia of the body, and, at the same time, it gives the intensity of the force. This is why bodies with different masses fall with the same acceleration. The force acting on the heavier one is bigger, but the heavier one is harder to accelerate. These two effects balance each other out. In the absence of air resistance, every object falls with an acceleration that depends only on the mass of the Earth and on the object's distance from its center.

***See:* Acceleration, Equation, Force, Mass, Measurement, Orbit, Relativity**

$$F_g = G\frac{m_1 m_2}{r^2}$$

[Eq. 5] ***The magnitude of the gravitational force (F_g) between two masses m_1 and m_2 is proportional to the product of the two masses and inversely proportional to the square of the distance between them (r). The constant of proportionality is G, the Newton's constant. The force is larger for larger masses, smaller for larger separations.***

THINK ABOUT IT: *General Relativity*

General relativity is the current theory describing gravitational interactions. Early in the 20th century, this theory replaced Newton's theory because it could explain phenomena that Newton's theory could not. For example, we now know that a mass like the Sun not only affects the motion of the planets but that it can bend the trajectory of light coming from other stars. Since light has no mass, we would not know how to use Newton's law to describe this bending. Nevertheless, as often happens in physics, the new theory "incorporates" the former one, though the theoretical basis of each can be very different. If the gravitational fields of a particular system are not too intense, and if the velocities of the bodies involved are much smaller than the speed of light, then Newton's theory can still be used to describe the interactions between massive objects, in the sense that the theory provides a good approximation of the general relativity results.

READ ABOUT IT: *Dreams & Journeys Through Physics*

Dreams of a Final Theory [1992] by Steven Weinberg — This Nobel Prize-winning physicist addressed Congress regarding the importance of superconductors for studying natural forces. Advances in science cannot be made if experimentation—and funding—does not keep pace with advances in theory. Because public support of scientific research is essential, so too is a public understanding of science.

Einstein's Dreams [1993] by Alan Lightman — Lightman's playful and fictitious glimpse into what Einstein might have been dreaming of in 1905 as he formulated the theory of relativity consists of 30 essays. This popular science book is an extraordinarily creative endeavor.

Journey into Gravity and Spacetime [1990] by John Archibald Wheeler — Wheeler, a contemporary of Einstein and Bohr, coined the term "black hole." Here, he offers what is widely appreciated as a general relativity handbook for nonmathematicians.

TALK ABOUT IT: *Beyond Physics—Gravity & Self-Security*

Masses attract each other in proportion to their size and closeness ... How do people fall under the "influence" of other people? ... What does it mean to "regroup," "get away," "get some space"? ... What is the process by which we come to know ourselves and feel secure? ... How do we represent ourselves? ... How does the "distance" between the self and the representation of the self (what we feel inside versus what we show the world) influence the way we are treated? ...

***The Lamentation* [1302-1305] by Giotto**
Fresco
Scrovegni Chapel, Padua

On the walls of the tiny Scrovegni chapel in Padua, Italy, beneath a luminous blue ceiling covered in gold stars, Giotto painted a massive religious story on a divinely human scale. These frescoes depict episodes beginning with the Old Testament and ending with The Last Judgment. In this panel, Giotto's solemn figures are the embodiment of grief—even the angels fall heavily from the sky, too anguished to maintain flight. *Gravity* has several sources here: the emotional and figurative weight of the characters, the limited depth of the settings, and the significance of Giotto's decision to simplify the compositions, thereby increasing narrative accessibility. According to Francesca Flores d'Arcais' *Giotto*, this attempt at the "actualization and secularization of the sacred"[9] advanced the level of engagement between image and audience, and marked the start of "the era of painting in Western art."[10]

***The Denial of Death* [1973] by Ernest Becker**

This Pulitzer Prize-winning book is full of moving psychological insights. Becker's description of the fear of death as the motivating force behind human thought and action contrasts with the Freudian view of behavior as sexually-motivated. Is anything more suggestive of the oppressive effects of *gravity* than the "denial of death," an immobilizing force that prevents self-actualization? Becker states that "modern man is drinking and drugging himself out of awareness, or he spends his time shopping, which is the same thing."[11]

HEAT

In spring the sun not only exerts an influence through the increased temperature of the air and earth, but its heat passes through ice a foot or more thick, and is reflected from the bottom in shallow water, and so also warms the water and melts the under side of the ice, at the same time that it is melting it more directly above, making it uneven, and causing the air bubbles which it contains to extend themselves upward and downward until it is completely honeycombed, and at last disappears suddenly in a single spring rain....

Every incident connected with the breaking up of the rivers and ponds and the settling of the weather is particularly interesting to us who live in a climate of so great extremes. When the warmer days come, they who dwell near the river hear the ice crack at night with a startling whoop as loud as artillery, as if its icy fetters were rent from end to end, and within a few days see it rapidly going out.

WALDEN AND OTHER WRITINGS
HENRY DAVID THOREAU

From: **HEAT** *n* (BEFORE 12C)
1 E (1) : ADDED ENERGY THAT CAUSES SUBSTANCES TO RISE IN TEMPERATURE, FUSE, EVAPORATE, EXPAND, OR UNDERGO ANY OF VARIOUS OTHER RELATED CHANGES, THAT FLOWS TO A BODY BY CONTACT WITH OR RADIATION FROM BODIES AT HIGHER TEMPERATURES, AND THAT CAN BE PRODUCED IN A BODY (2) : THE ENERGY ASSOCIATED WITH THE RANDOM MOTIONS OF THE MOLECULE, ATOMS, OR SMALLER STRUCTURAL UNITS OF WHICH MATTER IS COMPOSED F : APPEARANCE, CONDITION, OR COLOR OF A BODY AS INDICATING ITS TEMPERATURE

HEAT

When we put a cold and a hot body together, the hot body becomes less hot and the cold one less cold. If we leave them in contact for a long enough time, we can expect them to become equally warm or cold. When the temperatures of these two bodies have become the same, we say that the two bodies have reached *thermal equilibrium*. In the process, *heat* has been transferred from the hot body to the cold one. Because of this property of moving from one body to another, heat was once considered to be some sort of fluid.

It is now known that heat results from the *motion of molecules*. In a gas, molecules move randomly in every direction. In a hotter gas, the average speed of the molecules is larger than in a cooler one. The same can be said for liquids. And even in crystals, molecules move randomly about the equilibrium position that they occupy in the crystalline structure. *Temperature* is a measure of the average energy associated with molecular motion. In the example of the two bodies, the final state is the one in which the average velocity of the molecules of those bodies is the same. Microscopically, what has happened is that at the surface of contact, the fast-moving molecules of the hotter object increased the average motion of the molecules of the colder one. This "faster motion" then propagated slowly to the farther regions of the colder body. Note that *heat* is transferred, temperature is not.

Heat and temperature are two different quantities. Two bodies that are of different sizes but which are made of the same substance at the same temperature can transfer different amounts of heat to a colder body. A large hot body can transfer more heat to a colder body than a smaller body at the same initial temperature. In other words, in order to know *how much heat a body is able to transfer* to a colder one, we need to know its *volume*, although we do not need to know its volume in order to measure its temperature. The temperature is a quantity that can be measured in any small portion of the body. Heat transfer is measured in calories. A calorie is defined as the amount of heat necessary to increase the temperature of a liter of water by 1 degree Celsius.

***See:* Energy, Entropy, Friction, Motion, Particle**

THINK ABOUT IT: *First Laws of Thermodynamics*

When heat is transferred to a system or when work is done on it, its energy can increase. For example, we "do work" on a gas when we compress it—the volume of its container is reduced. On the other hand, the gas "does work" when it expands. The *first law of thermodynamics* says that the change in what is known as the *internal energy* of a system is equal to the sum of the heat we transfer to it and the work we do on it. The first law of thermodynamics is a law of conservation of energy. Heat transfer is energy transfer.

READ ABOUT IT: *Thermodynamics, Global Warming, Absolute Zero, & Entropy*

Boiling Point: How Politicians, Big Oil and Coal, Journalists, and Activists Have Fueled the Climate Crisis [2004] by Ross Gelbspan — Gelbspan sifts through years of science on climate change in order to shed light on the relationships between global warming, corporate excesses, and political corruption. He argues for an overriding plan of action to reverse the deleterious effects of greenhouse gases.

A Matter of Degrees: What Temperature Reveals About the Past and Future of Our Species, Planet, and Universe [2002] by Gino Segrè — Segrè gives a thorough overview of thermodynamics, global warming, *El Niño*, hydrothermal vents, absolute zero, and superconductivity. The author's uncle Emilio Segrè shared the Nobel Prize in physics in 1959 with Owen Chamberlain for their discovery of the anti-proton.

Warmth Disperses and Time Passes: The History of Heat [1998] by Hans Christian von Baeyer — Von Baeyer, the Chancellor Professor of Physics at the College of William and Mary, writes about efficiency, irreversibility, probability, entropy, and thermodynamics through the stories of scientists who made significant discoveries in the field.

TALK ABOUT IT: *Beyond Physics—Heat & Catalysts*

Consider social movements in American history (democracy, the abolition of slavery, organized labor, women's right to vote, civil rights) ... Who were the catalysts for new ideas? Who are they today? ... How might leaders maintain the interest of their constituents during the delays that are often intrinsic to change? ... The problem of "reach" ... How do original voices reach audiences? ... How important is it for citizens of a democracy to have a multiplicity of perspectives? ... Are there enough independent media systems in contemporary society to aid in the emergence of new "catalysts"? ...

***The Repentant Magdalen* [circa 1640] by Georges de La Tour**
Oil on canvas
National Gallery of Art, Washington, D.C.

La Tour painted Mary Magdalene, the woman described in the Bible as having been transformed through her faith in Christ. It is not known whether Mary Magdalene's "penitence" was actually a consequence of her failings—attributed at various times throughout history to immorality, vanity, materialism—or instead, a consequence of a decision by the Church to ascribe penitence to her. It is enough to appreciate the subject of La Tour's painting as a person caught in a moment of conversion. She is seated in contemplative solitude with the *heat* from a candle's flame warming her face while at the same time the fire draws back into surrounding darkness. It is the expert application of *chiaroscuro*—the interplay of dark and light without regard to color—that dramatizes what might most appropriately be described as a story of devotion and choice.

***Walden and Other Writings* [1854] by Henry David Thoreau**

Thoreau was the Harvard graduate who stepped out of society for a period of time to live in a cabin he built for himself in New England. Forsaking material wealth and comfort for the serenity of a simpler existence, this naturalist and individualist recorded his acquaintance with nature. This passage about the *heat* of spring is typical of the poetic descriptiveness of the author's Walden journals.

IMAGE 16

I write to show myself
showing people who show me my own showing.

WOMAN, NATIVE, OTHER
TRINH T. MINH-HA

Where the hunter-gatherer fills at most
one or two informal roles out of only several
available, his literate counterpart in an industrial
society must choose ten or more out of thousands,
and replace one set with another at different
periods of his life or even at different times of the day.
Furthermore, each occupation—the physician, the judge,
the teacher, the waitress—is played just so, regardless
of the true workings of the mind behind the persona.
Significant deviations in performance are interpreted
by others as a sign of mental incapacity or unreliability.
Daily life is a compromised blend of posturing
for the sake of role-playing and of varying degrees
of self-revelation. Under these stressful conditions,
even the "true" self cannot be precisely defined.

ON HUMAN NATURE
EDWARD O. WILSON

From: **IM-AGE** *n* (13C)
1 : A REPRODUCTION OR IMITATION OF THE FORM OF A PERSON OR THING 2 A : THE OPTICAL COUNTERPART OF AN OBJECT PRODUCED BY AN OPTICAL DEVICE (AS A LENS OR MIRROR) OR AN ELECTRONIC DEVICE B : A LIKENESS OF AN OBJECT PRODUCED ON A PHOTOGRAPHIC MATERIAL 3 A : EXACT LIKENESS B : A PERSON STRIKINGLY LIKE ANOTHER PERSON 4 A : A TANGIBLE OR VISIBLE REPRESENTATION 5 A (1) : A MENTAL PICTURE OF SOMETHING NOT ACTUALLY PRESENT : IMPRESSION (2) : A MENTAL CONCEPTION HELD IN COMMON BY MEMBERS OF A GROUP AND SYMBOLIC OF A BASIC ATTITUDE AND ORIENTATION B : IDEA, CONCEPT 6 : A VIVID OR GRAPHIC REPRESENTATION OR DESCRIPTION

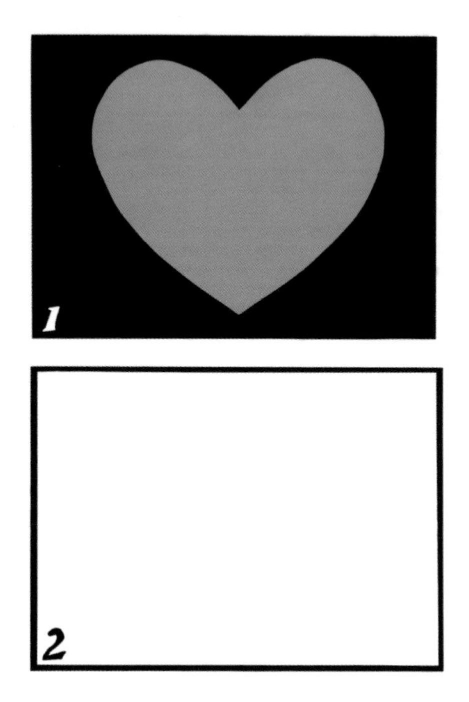

IMAGE

We see objects around us because they reflect light and because the reflected light beams travel along straight lines to our eyes. As we watch a landscape, light rays are reaching our eyes simultaneously from every point with different intensities and frequencies, corresponding to different luminosities and colors. The *image* of an object on a television or computer screen is generated by the emission of light, point by point—or, more properly, pixel by pixel. If we could slow down the action on the television screen, we would see that only one pixel is lit at any given instant in time, and then we would see the next one lit, and the next one lit, in sequence, until the whole screen was covered. If the pixels were lit in sequence fast enough, we would see them lit all at the same time, and we would see an image.

Mirrors and lenses bend light rays in a coherent way and can reproduce, magnify, or deform images of objects. In order to describe these processes, it is not necessary to know the physical nature of light. It is enough to know that light beams travel along straight-line paths unless they are reflected or refracted at the interface with a different medium, for example, at the interface between air and water or between air and glass. We can see the image of a mountain behind a lake on the lake's surface because the surface reflects light coherently when the water is still: light changes direction at the interface. Lenses also bend light rays at their surfaces. As the light beam enters the lens, the beam propagates in a different direction with respect to the incoming beam. Such deviation in the direction of propagation—as light *enters* a different material—is called refraction. The specific shape of the lens or the mirror can make parallel rays converge or diverge, as happens in optical devices such as telescopes, eyeglasses, or our eyes.

The laws of reflection and refraction are simple geometrical relations between different directions of propagation of light. These can be described in terms of angles, namely those between the surface (or the imaginary line perpendicular to the surface) and the direction of the incoming reflected and refracted rays. The branch of physics that studies optical images is therefore called *geometrical* optics.

See: Light, Reflection, Refraction

THINK ABOUT IT: *Motion Pictures & Television*

Television images do not move; they give the *impression of motion*. They are created by the sequential lighting of pixels. Motion pictures also give the impression of motion by showing a fast sequence of still frames of film. The frequency with which the frames are displayed is important because just as our eyes cannot distinguish very small spatial separations, they also cannot perceive differences in light over small separations in time.

READ ABOUT IT: *Politics of Representation, Surrealism, & Images of Absent Things*

Media Worlds: Anthropology on New Terrain [2002] by Faye Ginsburg, Lila Abu-Lughod, and Brian Larkin — Can humans ever speak *to* and *of* each other responsibly through imagery? This collection of essays on the ethics and politics of representation, and the effect of visual messages in an era of gargantuan media machines discusses the lowering standards of truth and fairness.

This Is Not a Pipe [1973] by Michel Foucault — How does language affect perception? The brilliant French philosopher's light-hearted essay on the nature of words and objects uses surrealist painter René Magritte's painting *Ceci n'est pas une pipe* [1926] as a centerpiece for analysis. The simple, dry solidity of the image of the pipe is undermined by Magritte's written text: *This is not a pipe*.

Ways of Seeing [1972] by John Berger — This extraordinary look into the darker side of art and advertising is based on a BBC television series of the same name. Berger discusses everything from images being used to conjure appearances of absent things to the dubious patronage of Renaissance artists by Italian merchants to the technical conventions that resulted from the special properties of oil paint.

TALK ABOUT IT: *Beyond Physics—Image & Falsehood*

If an image is "a reproduction or an imitation of a person or thing," how realistic can images in art, advertising, or news ever be? ... If an image is unrealistic, does it matter? ... Is a less realistic image more "truthful" in the admission of its falseness? ... Do movie/television images reflect things that are "there"? ... Where is "there"? ... Consider images of things most people will never actually see—the ocean floor, the Moon, Humphrey Bogart ... Would it matter if these things did not really exist? ... Informed spectatorship means becoming familiar with the intentions behind image production ...

***The Falling Cow Panel* [circa 17,000 B.C.E.]**
at The Cave of Lascaux
The Caves of Lascaux, France
(Montignac-Lascaux has a replica situated near the original)

Humans have been recording common experiences through *imagery* since prehistoric times. The cave paintings are among the few known man-made images that were largely—or entirely—unaffected by other images. *The Falling Cow Panel,* discovered in ancient caves in Lascaux, France, in 1940, shows a cow seeming to fall from the sky to the ground. Though the original meaning and function of such images is not known (they might have played some part in hunting or religious rituals), we can marvel at the fullness of expression, the maturity of layering, and the familiarity of the "narrative." This *image* bears a remarkable likeness to those made for modern children depicting "the cow jumping over the Moon."

***Woman, Native, Other* [1989] by Trinh T. Minh-ha**

In this original assessment of the effects of stereotype, Minh-ha describes objectification from the inside. She leads her reader to an understanding of *image* and identity through language that is personal, innovative, and audacious.

***On Human Nature* [1978] by Edward O. Wilson**

This is an excellent work by the Pulitzer Prize-winning zoologist (*The Ants,* 1990). Briefly-written chapters on "Sex," "Aggression," and "Altruism" read like essays that join to form a much broader view of the human *image.* Wilson investigates matters such as whether smiles are "hardwired," whether people are innately aggressive, and whether binary (i.e., either/or) classification is "normal."

LIGHT

LETTER 2, HELOISE TO ABELARD

You are the sole cause of my sorrow, and you alone can grant me the grace of consolation. You alone have the power to make me sad, to bring me happiness or comfort; you alone have so great a debt to repay me, particularly now when I have carried out all your orders so implicitly that when I was powerless to oppose you in anything, I found strength at your command to destroy myself. I did more, strange to say—my love rose to such heights of madness that it robbed itself of what it most desired beyond hope of recovery, when immediately at your bidding we changed my clothing along with my mind, in order to prove you the sole possessor of my body and my will alike.

LETTER 3, ABELARD TO HELOISE

Come too, my inseparable companion, and join me in thanksgiving, you who were made my partner both in guilt and in grace. For the Lord is not unmindful also of your own salvation, indeed, he has you much in mind, for by a kind of holy presage of his name he marked you out to be especially his when he named you Heloise, after his own name, Elohim. In his mercy, I say, he intended to provide for two people in one, the two whom the devil sought to destroy in one: since a short while before this happening he had bound us together by the indissoluable bond of the marriage sacrament. At the time I desired to keep you whom I loved beyond measure for myself alone forever, but he was already planning to use this opportunity for our joint conversion to himself.

THE LETTERS OF ABELARD AND HELOISE
TRANSLATED BY BETTY RADICE

From: **LIGHT** *n* (BEFORE 12C)
1 A : SOMETHING THAT MAKES VISION POSSIBLE B : THE SENSATION AROUSED BY STIMULATION OF THE VISUAL RECEPTORS C : AN ELECTROMAGNETIC RADIATION IN THE WAVE-LENGTH RANGE INCLUDING INFRARED, VISIBLE, ULTRAVIOLET, AND X-RAYS AND TRAVELING IN A VACUUM WITH THE SPEED OF ABOUT 186,281 MILES (300,000 KM) PER SECOND 5 A : SPIRITUAL ILLUMINATION 8 : SOMETHING THAT ENLIGHTENS OR INFORMS

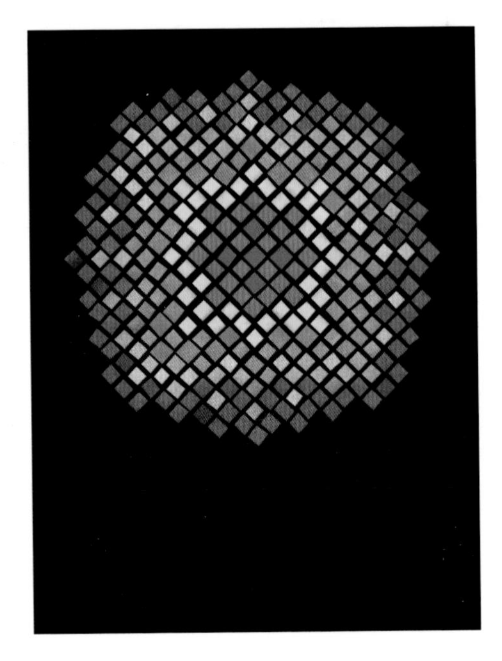

LIGHT

EARLY IN THE 19TH CENTURY, several discoveries were made regarding electric and magnetic phenomena. It became clear at the time that moving electric charges create a magnetic field and that a changing magnetic field creates an electric field.

By the middle of the 19th century, *James Clerk Maxwell* [1831-1879] unified the empirical laws of electricity and magnetism into a simple mathematical framework consisting of four basic equations. He predicted that an oscillating electric field generates an oscillating magnetic field, which, in turn, generates an oscillating electric field, and so on. This behavior results in a self-sustaining wave propagating at about 300,000 KM per second—exactly the speed of light. Interestingly, this prediction did not arise from direct experiments with light. *Light* was then recognized to be electromagnetic waves with frequencies that our eyes can perceive. Its propagation is similar to the propagation of water waves, though in the case of water, the waves consist of the up-down motion of water molecules, whereas, in the case of electromagnetic waves, it is the value of the electric and magnetic fields that oscillates. Unlike water, light does not need a medium to propagate. What propagates is a perturbation of the electromagnetic field. All known properties of light in Maxwell's time could be explained by the laws of electromagnetism.

By the beginning of the 20th century, the theory began to reveal its limits. From Maxwell's theory one could expect that an electromagnetic wave, consisting of electric and magnetic fields, would interact with electric charges. New experiments showed that the way energy is transferred from the wave to electrons in atoms was not the one predicted by Maxwell. Light, and electromagnetic waves in general, appeared to behave as particles. For example, electromagnetic waves of a given frequency transferred energy to electrons in small, discrete amounts. Each of these "particles" of light, or *photons*, carries an amount of energy proportional to the frequency of the wave. The theory of quantum mechanics, developed to make sense of this behavior, did not provide a new theory of electromagnetism, but a new way to describe the interactions of particles and fields at very small scales.

See: Reflection, Refraction, Wave, Wave/Particle Duality

THINK ABOUT IT: *Electromagnetic Waves*

The frequency of an electromagnetic wave is a measure of how many times the electromagnetic field *oscillates* in a given unit of time. Our eyes can detect only a limited range of frequencies. Radio waves, microwaves, and infrared radiation are electromagnetic waves with frequencies *lower than the visible range*. Ultraviolet radiation, X-rays, and gamma rays are electromagnetic waves with frequencies *higher than the visible range*. Gamma rays are the most energetic electromagnetic waves and are emitted in nuclear reactions. When we compare the energy of electromagnetic waves at different frequencies, what we are actually comparing is the energy of single photons. The total energy in a beam of light depends on the number of photons in the beam or, in terms of Maxwell's theory, on its intensity.

READ ABOUT IT: *Connections in the History of Ideas & Quantum Electrodynamics*

Connections [1978] by James Burke — In this completely original description of interconnecting events in history, Burke describes instances in which ordinary people have accidentally contributed to scientific advancement. The author has done a meticulous job in terms of research, and each page is filled with solid information concerning inventions ranging from the rocket to the television.

QED: The Strange Theory of Light and Matter [1985] by Richard Feynman — Feynman was not only a brilliant physicist and a fascinating lecturer, but a prolific author as well. This pocket-sized book is a good starting place for anyone interested in learning about his work and ideas. Feynman moves from Newtonian physics to quantum electrodynamics to basic interactions between light and electrons—all in a warm, conversational tone.

TALK ABOUT IT: *Beyond Physics—Light & Truth*

How is "seeking" light a metaphor for finding truth? ... How can truth be "found"? ... What might the process of truth discovery entail? ... If the answers that we find depend upon the questions we ask, is it possible to find truth that is untouched by personal bias? ... Is there a way to separate personal truth/individual perception of the world (like a particle) from shared truth/cultural perception of the world (like a wave)? ... How does the search for truth involve giving up biases? ... Is there one absolute truth? ... Who would determine that truth if there were? ... Government? Religion? Family? ... How does mutual friendship help us find "truth"? ...

An Angel [1978] by Marc Chagall
Stained Glass
All Saints Church, Tudeley, Kent, U.K.

Among the substances used by artists, few are as evocative of spirituality as stained glass. When natural *light* strikes the panels—each of which tells part of a story—the characters come to life. The passing radiance makes both vision and narrative possible. Marc Chagall, a Russian Jewish émigré, made these windows over the course of 15 years as a tribute to the daughter of Sir Henry and Lady d'Avigdor-Goldsmid, who died in a boating accident. The mother and daughter had once admired Chagall's stained glass during a trip to Jerusalem.

The Letters of Abelard and Heloise [1133-1138]
Translated by Betty Radice

Whether moved by the *light* of love or by the light of God, Heloise and Abelard devoted themselves to living passionately. Their correspondence reveals the tragic outcome of their entanglement, the complexity of their desire, and the depth of their respect for one another. Abelard, master of the Cathedral School of Notre Dame, and young Heloise were secretly married after having a love affair that produced a child. When Heloise's disgraced uncle tried to make the marriage public, Abelard hid Heloise in a convent. The uncle took revenge on Abelard by having him castrated, and the lovers retreated to monastery and convent, respectively.

MASS 18

The theory of relativity deduces, from its fundamental assumption, a clear and convincing answer to this question, an answer again of a quantitative character: all energy resists change of motion: all energy behaves like matter: a piece of iron weighs more when red-hot than when cool; radiation traveling through space and emitted from the sun contains energy and therefore has mass; the sun and all radiating stars lose mass by emitting radiation. This conclusion, quite general in character, is an important achievement of the theory of relativity and fits all facts upon which it has been tested.

Classical physics introduced two substances: matter and energy. The first had weight, but the second was weightless. In classical physics we had two conservation laws: one for matter, the other for energy. We have already asked whether modern physics still holds this view of two substances and the two conservation laws. The answer is: "No." According to the theory of relativity, there is no essential distinction between mass and energy. Energy has mass and mass represents energy. Instead of two conservation laws, we have only one, that of mass-energy. This new view provided very successful and fruitful in the further development of physics.

THE EVOLUTION OF PHYSICS
ALBERT EINSTEIN AND LEOPOLD INFELD

From: **MASS** *n* (15C)
1 C : THE PROPERTY OF A BODY THAT IS A MEASURE OF ITS INERTIA AND THAT IS COMMONLY TAKEN AS A MEASURE OF THE AMOUNT OF MATERIAL IT CONTAINS AND CAUSES IT TO HAVE WEIGHT IN A GRAVITATIONAL FIELD 2 : A LARGE QUANTITY, AMOUNT, OR NUMBER 3 A : A LARGE BODY OF PERSONS IN A COMPACT GROUP : A BODY OF PERSONS REGARDED AS AN AGGREGATE B: THE GREAT BODY OF THE PEOPLE AS CONTRASTED WITH THE ELITE—OFTEN USED IN PLURAL

MASS

MASS IS COMMONLY UNDERSTOOD AS the measure of the amount of matter contained in a body, but in physics, the second law of dynamics defines mass as a measure of the inertia of a body; namely, its tendency to oppose a change in velocity under the action of an external force. We measure the mass of an object by measuring how hard it is to set it in motion. Given a certain force, a body with a greater mass will experience a smaller acceleration than a body with a smaller mass. So, we can find the mass once we know the force we applied and the corresponding acceleration. The mass defined in this way is properly called the *inertial mass* or just the inertia of the body.

Commonly, matter is measured in terms of weight. In a grocery store, potatoes are sold by the pound. What is actually being measured is a force; specifically, the weight of a potato is the intensity of the gravitational force acting on it. On a spring balance, we read the amount of stretching of the spring: this gives a measure of the elastic force that opposes the gravitational force acting on the potato. Since the two forces must be equal in magnitude, we can then find the gravitational force. We do not measure any acceleration: once in equilibrium, the potatoes do not move. Therefore, we are not measuring their *inertial mass* as defined above; we are measuring their gravitational mass. The gravitational force is the only fundamental force in nature whose intensity depends on the mass of the object on which it acts. Precisely, the universal law of gravity states that the force of gravity between two objects—in this case, the Earth and the potato—is proportional to the masses of the objects and inversely proportional to the distance between them. In principle, the mass that appears in the law of gravity does not have to be the same mass that is defined by the second law of dynamics. In physics, every quantity is defined by the fundamental equation in which it appears. Mass appears in two different equations, so it is necessary to distinguish between two different properties of an object: its *inertial mass* and its *gravitational mass*, even though in experiments these properties have always turned out to correspond to the same value. Assuming them to be exactly the same property of a body is what led Einstein to a completely new theory of gravitation: *general relativity*.

***See:* Acceleration, Energy, Force, Gravity, Momentum**

$$F = m\,a \Leftrightarrow m = \frac{F}{a}$$

[Eq. 6] ***Newton's second law expressed in terms of the magnitudes of force and acceleration. The (inertial) mass of a body can be found by dividing the force acting on it by the acceleration it experiences.***

THINK ABOUT IT: *Mass and Special Relativity*

Prior to the 20th century, the definition of mass as a measure of inertia was not far from the idea of mass as a measure of an "amount of matter." But, in 1905, the special theory of relativity introduced different laws of dynamics. The definition of mass as a measure of inertia remained, but, according to the new theory, the energy of an object due to its state of motion—its *kinetic energy*—also contributes to its inertia. $E = mc^2$ means not only that a given mass corresponds to a certain amount of energy but that the inertia—the *mass*—of a particle is given by its energy. For objects moving at speeds much lower than the speed of light, the contribution of their kinetic energy to inertia is so small that it can be ignored, so Newton's law works well. For particles moving at speeds close to the speed of light, the contribution of their kinetic energy to their inertia cannot be ignored. The faster a particle moves, the harder it is to accelerate it further.

READ ABOUT IT: *Urban Space, Physics and Philosophy, & Climbing Mount Everest*

The City in History: Its Origins, Its Transformations, and Its Prospects [1961] by Lewis Mumford — This dense and beautifully researched classic contains findings on the history, culture, and development of cities. It moves from city as sanctuary and stronghold to city as suburb and slum to city as megalopolis and global center.

Concepts of Mass in Contemporary Physics and Philosophy [1961] by Max Jammer — Jammer uses mass as the groundwork for a philosophically-tempered history of physics. Though the book covers advanced material, it contains some exceptionally clear passages.

Everest [1998] by MacGillivray Freeman Films — This account of a journey to the world's tallest peak features the team who made the expedition, including the Sherpas who carried the film equipment to the summit. It chronicles the history of exploration on Mount Everest.

TALK ABOUT IT: *Beyond Physics—Mass & Persuasion*

In science, mass is a measure of resistance to movement ... What might establish personal mass? ... Can one's "inertia" be increased? ... On what foundations might we build a resistance to excessive persuasion? ... What does it mean to be "grounded"? ... How can a wide variety of friends and influences—art, music, literature, math, science, history, sports—help to stabilize people? ... Is it possible to become resistant to persuasion but remain receptive to beauty? ...

***Pietà* [1498-1499] by Michelangelo Buonarroti**
Marble
St. Peter's Basilica, Vatican City

There is no clearer example of the concept of *mass* in the arts than sculpture, and perhaps no finer example of sculpture than the *Pietà*, completed when Michelangelo was 23 years old. One has only to glance at the deep creases in the draped clothing and the tender weight of Jesus' body in his mother's arms to be moved irrevocably by the artist's vision and mastery. That this began as a formless piece of marble and ended as a work of enduring poignancy is testament not only to the artist and to the Renaissance but to the spirit of humanity, regardless of religious affiliation. In order to gain insight into Michelangelo's incredible skill, look at his unfinished sculptures—the *Slave* sculptures stored in the Galleria dell'Accademia in Florence and the *Pietà Rondanini* [1552-1564] in Milan's Sforza Castle.

***The Evolution of Physics* [1938]**
by Albert Einstein and Leopold Infeld

In his preface, Infeld describes Einstein as "the greatest scientist and the kindest man who ever lived."[12] The compliment seems well deserved: the book's objective—to teach physics to nonphysicists—is met in this accessible primary-source text. And what better place to read about *mass*? Likening the story of the universe to the greatest unsolved mystery, the authors lead the reader through the fundamentals of physics from Galileo to quantum mechanics.

MEASUREMENT 19

The traditional societies—whose representations of time are so difficult to grasp precisely because they are expressed in symbols and rituals whose profound meaning sometimes remains inaccessible to us—the traditional societies conceive of man's temporal existence not only as an infinite repetition of certain archetypes and exemplary gestures but also as an eternal renewal. In symbols and rituals, the world is recreated periodically. The cosmogony is repeated at least once a year—the cosmogonic myth serves also as a model for a great number of actions: marriage, for example, or healing.

What is the meaning of all these myths and rites? Their central meaning is that the world is born, grows weary, perishes, and is born anew in a precipitate rhythm.

The year—or what is understood by this term—corresponds to the creation, duration, and destruction of a world, a cosmos. It is highly probable that this conception of the periodic creation and destruction of the world, although reinforced by the spectacle of the periodic death and resurrection of vegetation, is not a creation of agricultural societies. It is found in the myths of pre-agricultural societies and is in all likelihood a lunar conception. For the most evident periodicity is that of the moon.... The lunar rhythms always mark a "creation" (the new moon) followed by a "growth" (the full moon), and a diminution and "death" (the three moonless nights).

TIME AND ETERNITY IN INDIAN THOUGHT
MIRCEA ELIADE
FROM: MAN AND TIME
EDITED BY JOSEPH CAMPBELL

From: **MEA-SURE-MENT** *n* (1751)
1 : THE ACT OR PROCESS OF MEASURING 2 : A FIGURE, EXTENT, OR AMOUNT OBTAINED BY MEASURING : DIMENSION

MEASUREMENT

THE STARTING POINT OF SCIENTIFIC research is the observation of nature and the *measurement* of well-defined physical quantities such as distance, time, mass, or temperature. This means assigning to such quantities a number by comparing them to a fixed unit of measurement, such as the meter, the second, the kilogram, or the degree. A second step consists of formulating a mathematical rule—a law—that relates the measured quantities in a physical problem and that makes new predictions. The last step is to perform experiments to measure the predictions and check the theory. The accurate measurement of predicted quantities is the *only* test a scientific model has to undergo. No matter how counterintuitive a theory might seem, experiments are the final determination of its soundness. Of course, any disagreement between theory and experiment might suggest an entirely new perspective that could potentially lead to another theory.

Every measurement comes with an *uncertainty*. As technology develops, the uncertainty on the measurements of physical quantities is reduced, and theories are subjected to increasingly stringent tests.

At the beginning of the 19th century, the theory of light formulated by Newton was widely accepted by most of the scientific community. Newton's theory assumed light to be made of traveling particles, it explained the way in which light is reflected and refracted, and it gave many other properties of light that could be tested. The theory also made the prediction that light travels at higher speeds in water or glass than in air, though at the time *it was impossible to measure such small differences in the speed of light.* The uncertainty on the measurement of the speed of light was then extremely large. As soon as experiments became accurate enough, and measurements became more precise, Newton's prediction proved wrong: *the speed of light in glass or water was actually smaller than in air*. Newton's theory of light as traveling particles had to be abandoned. A new theory was then formulated, in which light was described as the *propagation of an electromagnetic wave*. The new theory could explain reflection and refraction just as well as it could describe the observed differences in speed when light travels in different substances.

See: Equation, Uncertainty

THINK ABOUT IT: *Modern Measurements*

The prediction of *gravitational waves* that was made by the theory of general relativity has never been tested. Gravitational waves can be produced by the collision of very massive stars in the universe. Their effect on the Earth consists of a deformation of space, a squeezing and stretching of physical distances. Such an effect is so small that so far no gravitational wave has been detected. Experiments currently being developed aim to measure changes in distances caused by such gravitational waves, and these experiments are expected to be able to measure variations smaller than the size of the nucleus of an atom over a distance of more than four kilometers.

READ ABOUT IT: *The History of Measurement, Game Programming, & Quantum Gravity*

Measure for Measure: The Story of Imperial, Metric, and Other Units [2003] by Alex Hebra — This charming history of types of measurement from antiquity to the present shows how humans have had to refine techniques to make measurements increasingly precise.

Physics for Game Developers [2002] by David Bourg — The importance of physics to the development of realistic video games is discussed in this resource for those interested in programming. It covers ballistics, aerodynamics, and the motion of bodies through fluids.

Three Roads to Quantum Gravity [2000] by Lee Smolin — This book grapples with questions that are as philosophical as they are physical, such as "What is time?" and "How do we describe a universe in which we are participants?" Smolin is optimistic about the quest for quantum gravity and the discovery of a quantum theory of space-time.

Why Nothing Can Travel Faster Than Light [1993] by Barry Zimmerman and David Zimmerman — In "Measuring the Universe," the authors discuss the infinitely small (a cell) and the infinitely large (the diameter of the Earth). Also of interest: "A World Without Friction" and "How Dense Is Matter Inside a Black Hole?"

TALK ABOUT IT: *Beyond Physics—Measurement & Types*

Consider the problem of people describing people ... How do character archetypes (hero/villain/damsel in distress) simplify narratives? ... Can such simplicity ever be meaningful? ... Who profits from such reductive readings of the world? ... Why is it difficult to default to stereotypes when interacting on a one-to-one basis? ...

The Grand Canal from Carità to the Bacino di S. Marco
[circa 1730] by Canaletto
Oil on canvas
Her Majesty Queen Elizabeth II, The Royal Collection

Canaletto painted landscapes with such astonishing accuracy that many believe he used optical ruling instruments to *measure* the geometry of urban space and structure. In this particular "view" painting (or *veduta*, which was something similar to a postcard), Venice appears substantial, orderly, and refined, a pleasant vision of near-photographic realism set beneath an ocean of celestial blue. The corridors between buildings fall into dynamic shadow, alluding to cool, intriguing alleyways populated by busy Venetians.

"Time and Eternity in Indian Thought" [1951] by Mircea Eliade
from *Man and Time*, Edited by Joseph Campbell

How do we describe the mystery of time—or *measure* it—when the description depends heavily upon the describer, that is, *the subject of the experience of time*? The essays in this book discuss mortality, eternity, coincidence (including *déjà vu*), consciousness, science, faith, culture, and religion from several intellectual perspectives. Joseph Campbell, one of the most respected scholars and teachers of our time, has a way of extracting points of clarity, depth, and resonance from complicated material. In the nonfiction book series, *The Masks of God*, Campbell compares common themes and motifs underlying imagery, story, and myth from cultures throughout history in an attempt to celebrate the spiritual unity of humankind. Also: look for *Joseph Campbell and the Power of Myth* on video or DVD, the extraordinary set of interviews of the late Campbell by PBS correspondent Bill Moyers.

MOMENTUM 20

Providence has not created mankind entirely independent or entirely free. It is true that around every man a fatal circle is traced, beyond which he cannot pass; but within the wide verge of that circle he is powerful and free: as it is with man, so with communities. The nations of our time cannot prevent the conditions of men from becoming equal; but it depends upon themselves whether the principle of equality is to lead them to servitude or freedom, to knowledge or barbarism, to prosperity or to wretchedness.

DEMOCRACY IN AMERICA
ALEXIS DE TOCQUEVILLE

The task of setting free one's gifts was a recognized labor in the ancient world. The Romans called a person's tutelar spirit his genius. In Greece it was called a daemon.... An abiding sense of gratitude moves a person to labor in the service of his daemon. The opposite is properly called narcissism. The narcissist feels his gifts come from himself. He works to display himself, not to suffer change. An age in which no one sacrifices to his genius or daemon is an age of narcissism. The "cult of genius" which we have seen in this century has nothing to do with the ancient cult. The public adoration of genius turns men and women into celebrities and cuts off all commerce with the guardian spirits.

THE GIFT
LEWIS HYDE

From: **MO-MEN-TUM** *n* (1610)
1 : A PROPERTY OF A MOVING BODY THAT THE BODY HAS BY VIRTUE OF ITS MASS AND MOTION AND THAT IS EQUAL TO THE PRODUCT OF THE BODY'S MASS AND VELOCITY; *BROADLY* : A PROPERTY OF A MOVING BODY THAT DETERMINES THE LENGTH OF TIME REQUIRED TO BRING IT TO REST WHEN UNDER THE ACTION OF A CONSTANT FORCE OR MOMENT 2 : STRENGTH OR FORCE GAINED BY MOTION OR THROUGH THE DEVELOPMENT OF EVENTS : IMPETUS

MOMENTUM

IN PHYSICS, THE *MOMENTUM* OF a body is the product of its *mass times its velocity*. If a body has a velocity double that of another body, the two can still have the same momentum if the second one has a mass double that of the first. In other words, the product of the two quantities is the same. Also, momentum is a vector: it has a magnitude and a direction, the latter being the direction of the velocity.

According to the *principle of inertia*, a body on which no force is acting will keep its velocity constant in magnitude and direction. What if we have two bodies and no external force? Let's consider two astronauts in outer space where the force of gravity and air resistance are negligible. The astronauts form an isolated system. Imagine them initially at rest, one with respect to the other. In general, we cannot say that their respective velocities will stay constant. Let's say, for instance, that one astronaut pushes the other away, exerting a force internal to the system. The two would begin to move in opposite directions, and they would experience a change in their velocities. Thus, their individual velocities are not conserved: they are not zero anymore. What is conserved is the *total momentum of the system*. Remember that the initial total momentum of the system was zero because the respective momenta of the astronauts was zero—zero being their velocities. The final total momentum is also zero because the two astronauts gained opposite momentum with respect to their initial position. The sum of two vectors with the same magnitude and opposite direction is zero. The two astronauts will not necessarily have the same speed: if one is lighter than the other, he will have a greater speed, so the product of mass times speed will give the same magnitude for the momentum as the other one.

The second law of dynamics, $F = m\,a$, can also be expressed in terms of the momentum of a system by saying that an external force is equal to the rate of change of the total momentum of the system. If the external force is zero, there is no change in momentum: *the momentum is conserved*. For a single body of given mass, the rate of change of momentum is equal to the product of its mass times its acceleration.

***See:* Acceleration, Force, Mass, Motion, Position, Vector, Velocity**

$$\vec{p} = m\,\vec{v}$$

[Eq. 7] ***The momentum of a body is equal to the product of its mass times its velocity. The momentum is larger for a larger mass, as well as for a larger velocity.***

THINK ABOUT IT: ***Momentum & Velocity***

Momentum is a physical quantity that in layman's terms could be called "amount of motion." The total momentum of a system of moving bodies is given by the sum of the individual momenta of each body. Each body makes a contribution according to its *mass* as well as its *velocity*. On the other hand, the sum of the velocities of different bodies does not have any particular physical meaning. For instance, the total momentum of the system given by a train with 10 cars is the sum of the momenta of the cars and is equal to the momentum of an object moving with the velocity of the train with a mass equal to the mass of the whole train: that is the momentum of the train. In principle, the sum of the velocities can also be computed: it would be equal to 10 times the velocity of the train, although such a velocity has no meaning in describing the motion of the train.

READ ABOUT IT: ***Capitalism, Science, & Society***

The Common Good [1998] by Noam Chomsky — This professor of philosophy and linguistics at MIT is a frank critic of globalization, a champion of intelligent and conscious living, and a straightforward advocate of a sustainable future. *The Common Good* consists of transcripts from a series of seven interviews in which Chomsky deftly and fearlessly addresses some of the negative consequences of capitalistic policies. The book includes lists of organizations worth supporting and practical things concerned readers can do.

The Turning Point [1982] by Fritjof Capra — This absorbing book addresses the most challenging problems of our time and encourages us to think critically about science, religion, and the media. The film version, *Mindwalk* [1990], a conversation held by a scientist, a politician, and a poet was directed by Capra's brother, Bernt.

TALK ABOUT IT: ***Beyond Physics—Momentum & Peace***

How do ideas attain a greater or lesser "amount of motion"? ... What is the ideology of peace? ... Can peace be seen as separate from war? ... If inertia in physics is an amount of "resistance to motion," how might inertia be defined in regard to activism—is it analogous to personal comfort? ... Can humans work in times of relative comfort to guarantee a sustainable peace? ... Do we have an obligation to keep the struggle for peace at the forefront of public policy as many believe we do for civil rights, environmentalism, democracy? ... How does a struggle for an ideal necessitate the absence of that ideal? ... What is the meaning of the phrase "Think globally, act locally"? ...

***Panel with Striding Lion* [circa 604-562 B.C.E.]**
by Unknown
Glazed brick, Neo-Babylonian, Nebuchadnezzar II,
Mesopotamia, Babylon (modern Hilah); Modern State of Iraq
The Metropolitan Museum of Art, New York

This glazed brick mosaic dates from the reign of Nebuchadnezzar II, during which time Neo-Babylonia became a country of tremendous wealth and splendor—primarily at the expense of its neighbors, including the Egyptians, the Phoenicians, and the people of Jerusalem, who were deported to Babylon and enslaved. *Striding Lion* is art with a *momentum* generated by the political events that preceded its life as art: the work can hardly be considered independent of the circumstances of its production. Not only was it made with riches derived through a series of military conquests, but it was made to celebrate the supremacy of the new Babylonian Empire.

***Democracy in America* [1835] by Alexis de Tocqueville**

French aristocrat Tocqueville spent two years studying America's laws, customs, and citizenry in an attempt to get to the essence of democracy. This masterful work of scholarship has made a significant contribution to the history of human ideas. In it, Tocqueville asks important questions about democracy, such as "How can the *momentum* of a state based on equality be maintained?"

***The Gift* [1979] by Lewis Hyde**

In this survey of cross-cultural gift-giving practices, Hyde discusses the essentially incompatible notions of art and commerce. Using an anthropological model, he explores the meaning of generosity and the obligation of givers and receivers to help a gift achieve its particular *momentum* (e.g., should a gift be kept by the recipient, or should it remain in circulation?). Hyde's assessment of property, bartering, and the creative act is beautifully written and highly recommended.

MOTION 21

She sat listening to the music.
It was a symphony of triumph. The notes flowed up, they spoke of rising and they were the rising itself, they were the essence and the form of upward motion, they seemed to embody every human act and thought that had ascent as its motive. It was a sunburst of sound, breaking out of hiding and spreading open. It had the freedom of release and the tension of purpose. It swept space clean, and left nothing but the joy of an unobstructed effort.

ATLAS SHRUGGED
AYN RAND

From: **MO-TION** *n* (14C)
1 A : AN ACT, PROCESS, OR INSTANCE OF CHANGING PLACE : MOVEMENT B : AN ACTIVE OR FUNCTIONING STATE OR CONDITION 2 : AN IMPULSE OR INCLINATION OF THE MIND OR WILL 3 A : A PROPOSAL FOR ACTION

MOTION

ARISTOTLE [384-322 B.C.E.] DEFINED PHYSICS AS THE science that studies "change." The most basic "change" we experience is the change in positions of objects, their *motion*. Aristotle distinguished changes in the *location* of objects from changes in *quantity* and *quality* of substances, such as those that occur in chemical reactions. For him, these phenomena were all the subjects of physics, and although today they are studied by different disciplines, like chemistry and biology, the fundamental processes explaining the events that happen in our world are still the subject of physics. Ultimately, these processes entail some form of motion.

The greatest achievements in physics have consisted in identifying a common explanation for different observed facts or events. The motion of planets and falling objects on the Earth has been reduced to the effect of a single force, described by one simple equation: *the law of universal gravitation*. The properties of gases have been explained in terms of the microscopic motion of their molecules and the collisions of those molecules. Concepts such as heat and temperature have been related to the vibrations of atoms, and, as the word *thermodynamics* suggests, motion is at the basis of any thermal process. Chemical reactions and chemical compounds have been shown to depend on the electromagnetic interactions of protons and electrons. The electromagnetic force was then recognized as a fundamental force of nature, and even the motion of light was described as an electromagnetic phenomenon. By the end of the 19th century, physicists believed that the laws of gravity and those of electromagnetism could explain everything "that moves" in our world.

In the 20th century, physicists realized that on very small scales the description of motion has to be different from the one given by Newton's laws of dynamics. A new mechanics for the "quantum world" was developed. At the same time, Einstein provided a description of gravitational interactions. His approach modified the basic notions of space and time. According to the general theory of relativity, not only do objects move in space, but space itself "moves," or, as physicists say, space is a "dynamical quantity."

***See:* Acceleration, Force, Heat, Position, Relativity, Space-time, Velocity**

THINK ABOUT IT: *Frames of Reference*

When we say that an object is moving, it is always implied that the object is moving with respect to something else—something that is not moving. For example, a car that is moving on the street is moving with respect to the buildings and the street. In physics, the street would be referred to as a *frame of reference*. A frame of reference is needed to describe the motion of any object. The careful examination of the way motion and physical phenomena are described in different frames of reference is at the basis of the discoveries made by *Galileo Galilei* [1564-1642], Newton, and Einstein.

READ ABOUT IT: *Schrödinger, Cells, History, & Music*

In Search of Schrödinger's Cat: Quantum Physics and Reality [1984] by John Gribbin — This highly-acclaimed general science book provides readers with a nontechnical introduction to the interesting and often surprising features of quantum mechanics.

The Lives of a Cell: Notes of a Biology Watcher [1974] by Lewis Thomas — This National Book Award-winner [1975] contains essays that first appeared in the *New England Journal of Medicine*. Thomas describes motion in terms of the symphony of life—the "flow" of organelles, the "tingle" of mitochondria, and the "breathing" Earth.

A People's History of the United States [1980] by Howard Zinn — From the first sentence to the last of this bold account of American history (told from the perspective of women, workers, African Americans, Native Americans, et al.), readers will feel immersed in a river of important information. Zinn speaks of culture-shaping "events" such as slavery, the Depression, and the Vietnam War.

Science & Music [1937] by Sir James Jeans — Want to know how vibrating strings and organ pipes make music, or how different tones join to form consonance and dissonance? This classic offers a physical analysis of sound, complete with diagrams of stretched strings.

TALK ABOUT IT: *Beyond Physics—Motion & Creativity*

An object in motion "moves with respect to something else" ... Consider creativity as motion ... Is the dominant culture the "frame of reference" for the production of art? ... Why are humans moved to express their experiences through art? ... Think about the "stillness" of paintings and the "movement" of music ... Conversely, "movement" in paintings and "stillness" (but not silence) in music ...

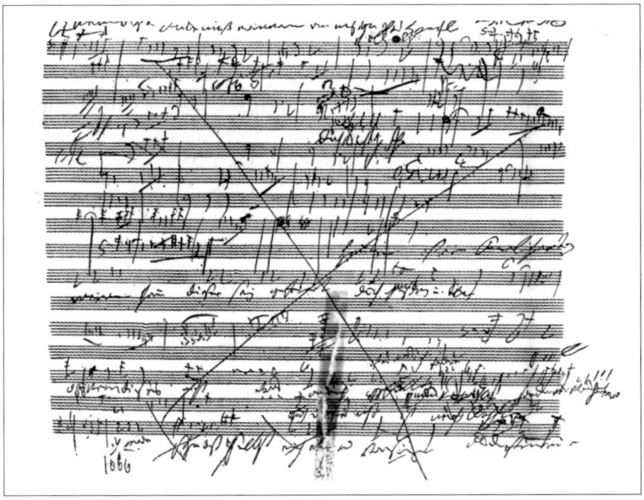

Ninth Symphony [1824] by Ludwig van Beethoven
Selection from the original manuscript
Private Collection

This extraordinary document is an amazing rendering of *motion*, especially considering that Beethoven was deaf when he wrote it. Imagine the notes coming to life in the still hush of the composer's mind, then being imprisoned on the neutral ground of paper and waiting—like little birds in little cages—to be set free time and again by musicians. The composer avoided social gatherings after experiencing the first signs of deafness at age 29, stopped playing in public at age 44, and went on to compose some of his greatest works (the piano sonatas, the string quartets, the *Missasolemnis*, and the Ninth Symphony) after the total loss of his hearing. It is said that he would put his ear to the floor and listen to the vibrations of his music, and that, following the debut of the Ninth Symphony, he didn't realize the audience was giving him an enthusiastic ovation because he was seated facing away from them in the front row. "Ode to Joy," the final movement of this piece, recently became the official musical theme of the European Union.

Atlas Shrugged [1957] by Ayn Rand

How to speak music? How to write sound? Rand, a Soviet émigré, a concerned American, a staunch capitalist, and an aesthetic idealist, in equal measure, writes of *motion* as it appears in the mind's eye—lyrically. Her description of "Halley's Concerto" is an extrapolation of a physical concept: the music ascends, breaks, hides, spreads, and sweeps space clean.

ORBIT 22

Here, as in many fairy tales, being pushed out of the home stands for having to become oneself. Self-realization requires leaving the orbit of the home, an excruciatingly painful experience fraught with many psychological dangers. This developmental process is inescapable; the pain of it is symbolized by the children's unhappiness about being forced to leave home. The psychological risks in the process, as always in fairy stories, are represented by the dangers the hero encounters on his travels.

THE USES OF ENCHANTMENT
BRUNO BETTELHEIM

A major clash between economics and ecology derives from the fact that nature is cyclical, whereas our industrial systems are linear. Our businesses take resources, transform them into products plus waste, and sell the products to consumers, who discard more waste when they have consumed the products. Sustainable patterns of production and consumption need to be cyclical, imitating the cyclical processes in nature. To achieve such cyclical patterns we need to fundamentally redesign our businesses and our economy.

THE WEB OF LIFE
FRITJOF CAPRA

From: **OR-BIT** *vt* (1943)
1 : TO REVOLVE IN AN ORBIT AROUND : CIRCLE 2 : TO SEND UP AND MAKE REVOLVE IN AN ORBIT; <~ A SATELLITE> ~*VI* : TO TRAVEL IN CIRCLES

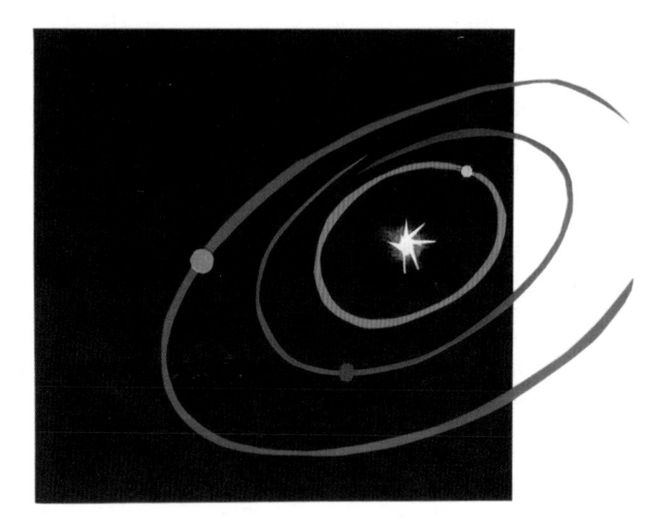

ORBIT

AN *ORBIT* IS THE PATH along which something continuously moves around something else, as the Earth orbits the Sun, and the Moon orbits the Earth. It was once thought that planets moved on perfectly circular paths around the Earth. This was because it was natural for philosophers of late antiquity and the Middle Ages to associate the motion of *celestial* bodies with the idea of *perfect* circles. When astronomers observed more carefully, they noted that the trajectory of the planets was *not exactly circular*, so they resorted to complicated combinations of circular paths to attempt to describe planetary motion. By the 16th century, observations became more accurate, and enough data had been amassed to show the limits of the existing theory. *Nicolaus Copernicus* [1473-1543] realized that by placing the Sun at the center of the solar system with all the planets orbiting around it, the description of their observed trajectories would be greatly simplified. Soon after, *Johannes Kepler* [1571-1630] proposed that the orbits were not circular, but elliptical. The Sun-centered theory formulated by Copernicus did not necessarily explain something the old one could not, but it was explaining it in a much simpler way, and this is an important feature for any scientific theory.

When we think of the story of Newton and the apple, we envision the apple falling straight to the ground. But imagine throwing an apple horizontally: it will make a curve before hitting the ground. Imagine throwing it faster: the faster we throw it, the larger the curve. Now imagine being on the highest mountain in the world and throwing the apple so fast that the curve it makes is larger than the curvature of the Earth—so large that the apple will not touch the ground at all. *We have put the apple in orbit.* Strictly speaking, the apple is still falling—it is still moving under the influence of gravity. From Newton's law, we can calculate the apple's trajectory: it is an ellipse. Every satellite orbiting Earth is constantly "falling." The Moon, too, is constantly "falling" on the Earth since it is always moving just under the influence of gravity—in space there is no air to slow it down—while its "initial" velocity does not allow it to "hit the ground." To fall, for a physicist, is not to hit the ground, but to move under the influence of gravity.

***See:* Acceleration, Force, Gravity**

THINK ABOUT IT: *Kepler's Laws*

I. Each planet moves in an ellipse with the Sun at one focus. [*The focus is not the center of the ellipse.*]
II. The line between the Sun and the planets sweeps over equal areas in equal time intervals. [*For this to be true, the planet has to move faster when close to the Sun, slower when far away.*]
III. The ratio of the cube of the semi-major axis to the square of the period of revolution is the same for each planet. [*This tells us how the size of the orbit is related to the period of revolution—the time it takes the planet to make a complete orbit.*]

READ ABOUT IT: *Astronomy & Space Travel*

Mapping the Sky: The Essential Guide to Astronomy [2001] by Leïla Haddad and Alain Cirou — This handsome book mixes photographs, diagrams of the solar system, drawings from antiquity (e.g., ones of the Moon made by Galileo), and star charts with descriptions of the history of cosmology. Suitable for students of stargazing.

Sojourner: An Insider's View of the Mars Pathfinder Mission [2003] by Andrew Mishkin — This interesting book by a senior member of NASA's robotics program offers a look into the Mars Pathfinder Mission. Mishkin discusses funding, staffing, design, and implementation for Sojourner, the first self-guided space exploration robot.

Cosmos [1979] produced by Adrian Malone — The landmark 13-part series hosted by Carl Sagan and based on his best-selling book of the same name has met with overwhelming success since first airing—over 500 million people in 60 countries have seen it. Sagan did not underestimate the intelligence of the viewers, and the program's broad acceptance shows general audiences to be more capable of handling tough material than is commonly believed. *Cosmos* won an Emmy, a Peabody, and a Hugo, all in 1981.

TALK ABOUT IT: *Beyond Physics—Orbit & Home*

The orbit of home ... What is home, and where? ... Species? Nation? Culture? Religion? Family? Hearth? ... What does it mean to say, "I was born here"? ... If home is the physical place where we are born, are we all exiles? ... How does building a story of home increase our sense of self-importance? ... Is the "story of you" just a link in a chain of events extending back through your ancestors and forward through your descendants? ... Is home simply a place where there is companionship and the self is at ease and intact? ...

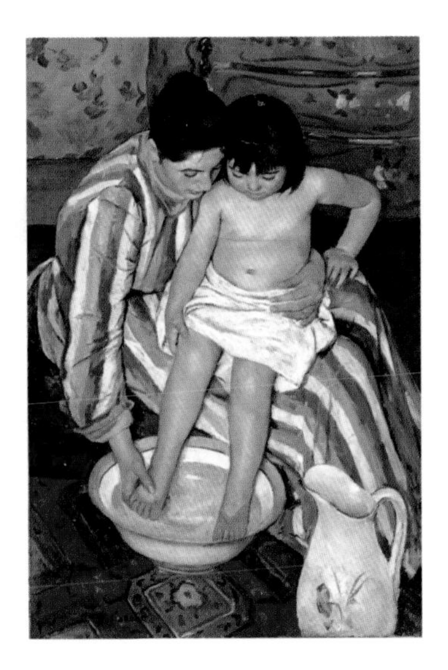

***The Bath* [1893] by Mary Cassatt**
Oil on canvas
The Art Institute of Chicago, Chicago

Cassatt was an American woman of independent means who studied art in Europe, befriended Edgar Degas, joined the Impressionists, and settled permanently in Paris. Because of her wealthy social contacts, she was instrumental in helping the Impressionists gain a following in America. The elevated vantage point in *The Bath* gives one the sense of encroaching on the intimate *orbit* of mother and child. The work has the feel of a captured moment, and it reflects the growing influence of observational realism (e.g., in photography) at the time. Realism, a technique that fell between Romanticism and Symbolism, was not expressive of heroic ideals, but of proximity and objectivity and very modern questions, such as, what constitutes "truth"?

***The Uses of Enchantment* [1976] by Bruno Bettelheim**

In this engrossing look into the importance of fantasy to the human psyche, Bettelheim offers fascinating reasons for the thematic prevalence of wicked stepmothers, animal-grooms, knights, simpletons, and of course, child characters forced to leave the *orbit* of home.

***The Web of Life* [1996] by Fritjof Capra**

Capra discusses the human condition in terms of wholeness versus fragmentation, and he stakes out society's battles between a linear, teleological world view, and a cyclical, (*orbiting*) nature-based view in this very important philosophical survey of modern science.

PARTICLE 23

Many years had elapsed during which nothing of Combray, except what lay in the theatre and the drama of my going to bed there, had any existence for me, when one day in winter, on my return home, my mother, seeing that I was cold, offered me some tea, a thing I did not ordinarily take. I declined at first, and then, for no particular reason, changed my mind. She sent for one of those squat, plump little cakes called "petites madeleines," which look as though they had been moulded in the fluted valve of a scallop shell.
And soon, mechanically, dispirited after a dreary day with the prospect of a depressing morrow, I raised to my lips a spoonful of the tea in which I had soaked a morsel of the cake. No sooner had the warm liquid mixed with the crumbs touched my palate than a shiver ran through me and I stopped, intent upon the extraordinary thing that was happening to me. An exquisite pleasure had invaded my senses, something isolated, detached, with no suggestion of its origin. And at once the vicissitudes of life had become indifferent to me, its disasters innocuous, its brevity illusory—this new sensation having had the effect, which love has, of filling me with a precious essence; or rather this essence was not in me, it was me. I had ceased now to feel mediocre, contingent, mortal. Whence could it have come to me, this all-powerful joy? I sensed that it was connected with the taste of the tea and the cake, but that it infinitely transcended those savours, could not, indeed, be of the same nature. Where did it come from? What did it mean? How could I seize and apprehend it?

SWANN'S WAY
FROM: À LA RECHERCHE DU TEMPS PERDU
[REMEMBRANCE OF THINGS PAST: VOLUME I]
MARCEL PROUST
TRANSLATED BY C. K. SCOTT MONCRIEFF

From: **PAR-TI-CLE** *n* (14C)
1 A : A MINUTE QUANTITY OR FRAGMENT B : A RELATIVELY SMALL OR THE SMALLEST DISCRETE PORTION OR AMOUNT OF SOMETHING 3 : ANY OF THE BASIC UNITS OF MATTER AND ENERGY (AS A MOLECULE, ATOM, PROTON, ELECTRON, OR PHOTON)

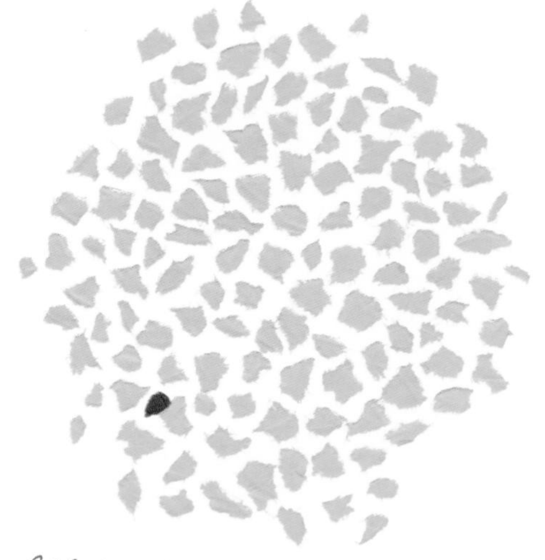
PARTICLE

THE WORD *PARTICLE* MEANS "LITTLE part." When we magnify something, we are able to see an internal structure that is not apparent to our eyes; for example, if we look at a drop of blood under a microscope, we'll see cells. Even after zooming in quite far, a great deal of structure remains to be seen. Cells are made of molecules, which are made of atoms. When we say that something is composed of atoms, we don't mean that it is "packed" with them. In fact, the forces that bind atoms also keep them at a certain distance from each other. The same is true within the atom, which is much bigger than its nucleus—approximately one hundred thousand times bigger—indicating that the atom is almost empty. What we feel when we touch something is actually the strength of the electromagnetic field holding the atoms together.

Atoms are made of electrons orbiting nuclei, and nuclei are made of protons and neutrons. These protons, neutrons, and electrons are called particles. Protons and neutrons are built of particles called quarks. No one has yet found structure inside quarks or electrons, and so they are considered "size-less" point particles. We do not know whether they actually are size-less, since any evidence of internal structure they might have has not been observed. We refer to them as elementary particles. The photon, a particle of light with zero mass, is also an elementary particle. Until the end of the 19th century, light was described as an electromagnetic wave. A wave consists of something oscillating, such as the water in the ocean. In the case of light, what is oscillating is the value of the electromagnetic field in a given point in space. At the beginning of the 20th century, experiments showed that, at very small scales, light may also behave like a particle. It was noticed that a beam of light was exchanging energy in small, discrete amounts. This behavior was explained by assuming that the light beam was made of photons, each carrying a fixed amount of energy, though at the time physicists were unable to reconcile this hypothesis with the successful electromagnetic wave theory. Now, quantum electrodynamics, which describes the interactions between photons and electrons, is one of the most successful theories of physics.

***See:* Light, Position, Wave/Particle Duality**

THINK ABOUT IT: *Dark Matter*

Although everything we see on our planet—and in stars and galaxies—is ultimately made of protons, electrons, photons, and a few other known elementary particles, the greater part of the matter in the universe (almost 90%) appears to be made of something else. Cosmologists refer to this unknown kind of matter as *dark matter*. It is dark because apparently it does not interact with the *electromagnetic field*. Dark matter does not reflect light. We know that it must be made of particles with a mass because we can measure its gravitational effects. But that is essentially all that is known about the most abundant particles in the universe.

READ ABOUT IT: *Atoms & Science Around the World*

Almost Everyone's Guide to Science: The Universe, Life and Everything [1998] by John Gribbin with Mary Gribbin — This enjoyable book for beginning scientists moves from the largest to the smallest components of existence ("Particles" to "Fields"), and covers everything from DNA to the Big Bang.

Atom: Journey Across the Subatomic Cosmos [1991] by Isaac Asimov — This Russian-born, Brooklyn-raised biochemist wrote more than 500 books before his death in 1992. In this, his last physics book, Asimov investigates the universe from the perspective of its smallest part: the atom. *Atom* is intended for an older audience than Asimov's classic and frequently updated *Inside the Atom* [1956], another excellent read, full of delightful diagrams and clear text.

The Rise of Early Modern Science: Islam, China, and the West [1993] by Toby E. Huff — Huff investigates various cultural stigmas (and values) attached to the pursuit of knowledge and the free flow of information. He also discusses advances in science in terms of the religious and political challenges that had to be overcome.

TALK ABOUT IT: *Beyond Physics—Particle & Loneliness*

Individuality versus loneliness ... Isolation by preference (hermits), punishment (prisoners), or profession (astronauts) ... When is solitude restorative and when is it a form of avoidance? ... How does wearing headphones in public allow us to block out our surroundings? ... Why is reading in public a more social alternative? ... What did philosopher Immanuel Kant mean by stating that people should use humanity not merely as a "means," but as an "end"? ... Is the individual obligated to remain connected to the community? ...

***A Sunday on La Grande Jatte—1884* [1884-1886]**
by Georges Seurat
Oil on canvas
The Art Institute of Chicago, Chicago

When he was 26 years old, Seurat was invited by fellow painter Claude Pissarro to join the final Impressionist exhibition in 1886. Seurat, who died soon afterward at the age of 31, created a stir with *La Grande Jatte*, which, with its maddeningly precise *particles* of paint, epitomizes a modernist reform of Impressionism referred to as Pointillism. Whereas the *mise-en-scène*—the middle class poised on the banks of the Seine—is hardly shocking, the labored and utterly conscious rendering of leisure results in an inescapable irony. Whether or not it was intentional, Seurat's painting contains a critique of middle-class culture, in which proximity influences perception: if one is far away from the scene, one will see ideally, but inaccurately; if one is close, one will see accurately—but *too* accurately. Marxist philosopher Ernst Bloch called the painting "a mosaic of boredom" and a "landscape of painted suicide."[13]

"Swann's Way" [1913] by Marcel Proust
from *À la recherche du temps perdu*
Translated by C. K. Scott Moncrieff

The memory of the scallop-shaped *petites madeleines* the narrator ate at his mother's home in Combray leads to a series of unexpected memories, like *particles*, which eventually are linked through narrative to reconstitute the wholeness of a personality. This painstaking recollection speaks to the power of personal reflection to help people (and artists) discover the singular meaning of life.

POSITION 24

Then he went into the enclosure and he looked carefully and he saw the place where he would lie below his father and his uncle and above Ching and not far from O-Lan. And he stared at the bit of earth where he was to lie and he saw himself in it and back in his own land forever. And he muttered, "I must see to the coffin."

THE GOOD EARTH
PEARL S. BUCK

My own sex, I hope, will excuse me, if I treat them like rational creatures, instead of flattering their fascinating graces, and viewing them as if they were in a state of perpetual childhood, unable to stand alone. I earnestly wish to point out in what true dignity and happiness consists. I wish to persuade women to endeavor to acquire strength, both of mind and body, and to convince them that the soft phrases, susceptibility of heart, delicacy of sentiment, and refinement of taste, are almost synonymous with epithets of weakness, and that those beings who are only the objects of pity, and that kind of love which has been termed its sister, will soon become objects of contempt.

A VINDICATION OF THE RIGHTS OF WOMAN
MARY WOLLSTONECRAFT

From: **PO-SI-TION** *n* (14C)
1 : AN ACT OF PLACING OR ARRANGING 2 : A POINT OF VIEW ADOPTED AND HELD TO 3 A : THE POINT OR AREA OCCUPIED BY A PHYSICAL OBJECT 5 A : RELATIVE PLACE, SITUATION, OR STANDING B : SOCIAL OR OFFICIAL RANK OR STATUS C : AN EMPLOYMENT FOR WHICH ONE HAS BEEN HIRED D : A SITUATION THAT CONFERS ADVANTAGE OR PREFERENCE

POSITION

WE CAN DESCRIBE WHERE SOMETHING is only with respect to something else. The position of a physical object is always given with respect to another object, which we call a *body of reference*. Given a body of reference, we can measure distances and orientations, and we can describe the *position* of an object in terms of a set of numbers called coordinates.

Conceivably, there are several ways to indicate the location of a treasure buried on an island. Our map might tell us to "walk 10 miles north, then 10 miles east." But without a starting point and without knowing where north is, the instructions on the map would be useless. We need an origin and a direction of reference. If we start from the main harbor, the harbor would become the origin of our coordinate system. We could then use a compass to find north. This map makes use of the Cartesian system of coordinates, named after the French mathematician and philosopher *René Descartes* [1596-1650]. A Cartesian system describes distances along perpendicular directions. We could get to the same treasure with a map that instructed us to "walk 14.14 miles northeast." These indications are again given in terms of two numbers. The first number (14.14 miles) is the total distance we have to walk, while the second is the angle implied by the word "northeast;" that is, an angle of 45 degrees with north direction. We would move then along a diagonal straight to the treasure. Again, we need to know the origin given by the harbor and the north direction given by the compass. This second document provides locations in terms of an angle and a distance, and makes use of a "polar" coordinate system as opposed to the Cartesian one where the two numbers are both distances.

In the examples above, the two coordinate systems use the same *frame of reference*. Also, in both cases the coordinates are two numbers since they describe positions on a surface, a two-dimensional space. For a generic position in a three-dimensional space, we would need three numbers, while for a one-dimensional space, one number is sufficient. The position of a train of a specific line of the subway can be given in terms of the distance from the first stop.

See:* *Angle, Measurement

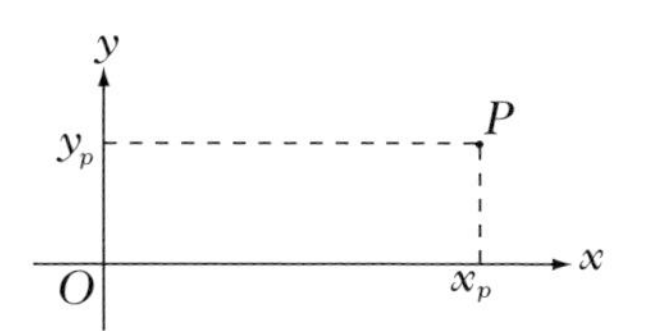

[Ill. 2] ***The position of a point P on a plane in Cartesian coordinates. The coordinates of the point are x_p and y_p, while x and y are the coordinates' axes. The origin of the frame of reference is O.***

THINK ABOUT IT: ***Points of Origin***

The choice of an *origin* and a *reference direction* defines the *frame of reference* used to describe physical phenomena. No frame of reference is more correct than any other, but some will be more convenient for a given system. By "convenient" we mean that the mathematical formulas describing a particular physical system in such a frame have a simpler form. For instance, the trajectory of Mars is an ellipse if the position of the Sun is chosen as the origin of the coordinate system, but that trajectory would be a terribly complicated geometrical figure if the position of the Earth were chosen as the origin instead.

READ ABOUT IT: ***Chronometers & Antique Maps***

The Illustrated Longitude [1998] by Dava Sobel and William Andrewes — Pendulums could not be counted on for accuracy at sea, so British clockmaker John Harrison built a chronometer, which used a "fourth temporal dimension" to link points on a three-dimensional globe. Supposedly, Newton didn't think it could be done. Was this the beginning of "space-time"?

Cartographica Extraordinaire: The Historical Map Transformed [2004] by David Rumsey and Edith M. Punt — These exquisite maps of North and South America from the 18th and 19th centuries have been re-printed from Rumsey's collection of 150,000 items—maps, atlases, and supporting materials. One especially remarkable piece is the *Cuadro histórico-geroglifico de la peregrinación de las tribus Aztecas que poblaron el Valle de Mexico* [1858], taken from the first national atlas to be published and printed in Mexico by a Mexican.

TALK ABOUT IT: ***Beyond Physics—Position & Place***

If, at any given moment, we occupy a physical location according to Cartesian coordinates (i.e., an "x, y, z"), what might be the location of our personal "position," our place in the world (i.e., our standard and manner of living)? ... Do we envision ourselves as belonging to the place we currently occupy or the place we hope to go? ... How do we define "here" and "there"? ... In which of the two places does happiness lie? ... How long will we be "here"? ... How do we intend to get "there"? ... Is a focus on the future a sign of careful planning, or can it constitute an escape from the present? ... From where/whom are the traditional definitions of success and progress derived? ... If human intention is focused too heavily on the future, do we risk putting ourselves in exile from the present? ... Choice versus chance: how much influence do people really have over events in their lives? ...

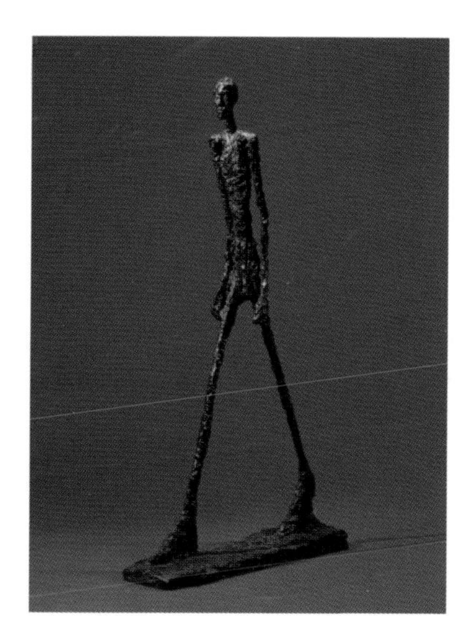

Walking Man II [1960] by Alberto Giacometti
Bronze
National Gallery of Art, Washington, D.C.

Giacometti's emaciated bronze figure is a character of harrowing loneliness and utter liberation. His *position* is as slender as a thread, as tiny as the point of a pin. He cannot be defined by possessions or associations. There is nothing external to him: he is himself, he is everyone. This post-modern exploration of "there" and "not there" invites us to confront the persistence of presence in the face of feelings of absence—an absence that is simply too vast to be relevant.

The Good Earth [1931] by Pearl S. Buck

Beneath the tale of rural tradition and culture in pre-Communist China lies an evaluation of the meaning of life in relation to work. Is it inevitable that our best intentions and endeavors be subsumed by the forces of nature, or is it possible to make contributions that are lasting? Buck, the first woman to win a Nobel Prize in Literature (she also won a Pulitzer), achieved the latter. She devoted her life to changing the *positions* of the underprivileged and the neglected.

A Vindication of the Rights of Woman [1792]
by Mary Wollstonecraft

Wollstonecraft mounted this treatise to feminism out of an abiding concern for the demeaning political and social *positions* held by women. She encouraged women to be feminine *and* "full and equal human subject[s]"[14] with strength of mind, body, and character.

POWER 25

Being envied is a solitary form of reassurance.
It depends precisely upon not sharing your
experience with those who envy you. You are
observed with interest but you do not observe with
interest—if you do, you will become less enviable.
In this respect the envied are like bureaucrats; the
more impersonal they are, the greater the illusion (for
themselves and for others) of their power. The power
of the glamorous resides in their supposed happiness:
the power of the bureaucrat in his supposed authority.
It is this which explains the absent,
unfocused look of so many glamour images. They
look out over the looks of envy which sustain them.

WAYS OF SEEING
JOHN BERGER

A prudent prince cannot and should not keep
his word when to do so would go against his interest, or
when the reasons that made him pledge it no longer apply.
Doubtless if all men were good, this rule would be bad; but
since they are a sad lot, and keep no faith with you, you
in your turn are under no obligation to keep it with them.
Besides, a prince will never lack for legitimate excuses
to explain away his breaches of faith. Modern history will
furnish innumerable examples of this behavior, showing
how many treaties and promises have been made null
and void by the faithlessness of princes, and how the man
succeeded best who knew best how to play the fox. But it
is a necessary part of this nature that you must conceal
it carefully; you must be a great liar and a hypocrite.
Men are so simple of mind, and so much
dominated by their immediate needs, that a deceitful man
will always find plenty who are ready to be deceived.

THE PRINCE
NICCOLÒ MACHIAVELLI

From: **POW-ER** *n* (13C)
1 A (1) : ABILITY TO ACT OR PRODUCE AN EFFECT 2 A : POSSESSION OF CONTROL, AUTHORITY, OR INFLUENCE OVER OTHERS 3 A : PHYSICAL MIGHT B : MENTAL OR MORAL EFFICACY C : POLITICAL CONTROL OR INFLUENCE 6 A : A SOURCE OR MEANS OF SUPPLYING ENERGY

POWER

POWER AND ENERGY ARE OFTEN thought to mean the same thing, but in physics the terms actually denote clearly distinct quantities. Power measures the *rate at which energy is used to do work or the rate at which it is transformed into another form of energy*. In order to perform a certain task, we need a certain amount of energy. But in order to perform the same task in a certain amount of time, we need enough power.

Climbing stairs requires energy—the *chemical energy* stored in our body. As we climb, most of this chemical energy is converted into *gravitational potential energy* because we move to a higher position. Suppose we walk up to the 10th floor of a building. The gravitational potential energy we gain depends exclusively on the height of the final position, not on the way we get to that position or how fast we get to it, and that energy is equal to the work our muscles do. The average power required to climb the stairs is equal to the total energy employed divided by the time it takes to reach the final position. What this means is that walking up to the 10th floor in 10 minutes requires half the power required to walk to the same floor in 5 minutes. The faster a force "does" a particular amount of work, the larger the power. The total energy needed to reach the 10th floor in both cases is the same—but the power is not.

Power is measured in *watts* while energy is measured in *joules*. One watt corresponds to 1 joule per second. A 60-watt designation written on a light bulb indicates that the bulb will consume 60 joules every second. It does not make sense to ask how much energy a given bulb consumes, since the answer depends on the time the bulb is left on. It makes better sense to ask how much energy the bulb consumes *per hour*, that is to say, *the power*. The energy consumed by a bulb, for the most part, is given away as *electromagnetic radiation*—or, light. The measure of the power consumed is also a measure of the rate at which the bulb gives away energy, namely its *luminosity*, or how bright it can become. As a matter of fact, luminosity is the word used to indicate the power of stars. The luminosity of a star is the amount of energy it radiates per unit time.

See: Energy, Force, Work

$$P = \frac{W}{\Delta t}$$

[Eq. 8] ***The average power (P) is equal to the work (W) done by a force divided by the time elapsed (Δt).***

THINK ABOUT IT: *Human Power*

While sleeping, a human being consumes approximately as much power as a light bulb, about 80 watts. This number measures the *rate at which we consume energy* to keep us warm. If we are not involved in any particular physical activity, the power we need while awake is about 100 watts, but it can increase 10 times during intense exercise.

READ ABOUT IT: *Robots, Atomic Bombs, & Creativity*

Flesh and Machines: How Robots Will Change Us [2002] by Rodney Brooks — The director of MIT's Artificial Intelligence Lab describes life on the inside of research and development, which depends as much on imagination as on advances in technology. A.I. is already integrated into human life via eye and ear implants, artificial limbs, and robots used for agriculture, surveillance, and space exploration.

The Making of the Atomic Bomb [1986] by Richard Rhodes — Rhodes chronicles the Manhattan Project and those who worked on it (Robert Oppenheimer, Niels Bohr). The Pulitzer Prize-winning book, which confronts the dark business of bomb-making, also won the National Book Award and the National Book Critics Circle Award [1987].

Creators on Creating [1997] by Frank Barron, Alfonso Montuori, Anthea Barron — This collection of writings on the creative process, taken from the writings of great musicians (e.g., Igor Stravinsky and Peter Tchaikovsky), authors, and scientists (e.g., Richard Feynman and Leonardo da Vinci), is inspirational. It contains great discussions of the importance of maintaining an open mind and heart.

TALK ABOUT IT: *Beyond Physics—Power of Persona*

What distinguishes a hero from a celebrity? ... Is celebrity attained through great deeds, uncommon strength, superior talent? ... From where does a celebrity's influence come? ... Do they possess power of persona (charm), of access (to media), or of ubiquity (every-whereness)? ... Are they singled out for fame by their "saleability"? ... How is fame a construct that helps the marketplace sell things? ... Who gets more press—Brad Pitt or Martin Luther King, Jr.? ... Paris Hilton or Eleanor Roosevelt? ... Would anyone buy a magazine with Susan B. Anthony on the cover? ... What do we gain by giving attention to celebrities? ... What do we lose? ... Do we feel less anonymous? ... As celebrities work to control the presentation of their personas, how does artificiality work against them, creating public demand for the "real" (e.g., "paparazzi" photographs)? ...

***Portrait of a Lady* [1460] by Rogier van der Weyden**
Oil on panel, painted surface
National Gallery of Art, Washington, D.C.

With her pronounced forehead, egg-shaped head, pale skin, and knowing look, this woman of the distant past appears remarkably like a contemporary version of a woman of the future. Who is she? What is she is considering? With each next feature—her grave and regal bearing, the serene purse of her lips, the solemn downward cast of her eyes, the determined grip of her hands—we are drawn further into intrigue. The artist exploits the darkness beautifully: there are no details beyond those articulated by the woman's *powerful* persona.

***Ways of Seeing* [1972] by John Berger**

This daring diatribe against the media industry alerts readers to the ways in which images are used by the *powerful* to captivate and oppress the powerless. Berger insists that we choose awareness over manipulation, describing such advertising schemes as the romantic use of nature and the unconscionable reliance on sex, stereotype, and wealth to arouse envy and self-loathing in viewers.

***The Prince* [1513] by Niccolò Machiavelli**
Translated by Robert Adams

This infamous political theorist has earned a reputation for counseling the use of *power* for power's sake. Yet, the text must be seen in relation to the violence and corruption that characterized the demise of the Italian state system. In the book, Machiavelli calls for an end to chaos.

PRESSURE 26

The more automatic our organizations become,
the more necessity there is for a system of regulation;
and that system, like the clock's, must be adjusted
in terms of an external standard, independent of
the mechanism. In the case of a clock—the revolution
of the earth: in the case of human institutions
—the whole nature of man, not just that portion of it
which has been fascinated by the machine and become
submissive to its needs. So with cities: to correct
the deficiencies of our over-mechanized civilization,
we shall have to build up a multi-centered
system of control, with a sufficient development
or morality, intelligence, and self-respect to be
able to arrest the automatic processes—mechanical,
bureaucratic, organizational—at any point where
human life is in danger or the human personality
is threatened with loss of values and choices.

THE CITY IN HISTORY
LEWIS MUMFORD

From: **PRES-SURE** *n* (14C)
1 A : THE BURDEN OF PHYSICAL OR MENTAL DISTRESS B : THE CONSTRAINT OF CIRCUMSTANCE : THE WEIGHT OF SOCIAL OR ECONOMIC IMPOSITION 2 : THE APPLICATION OF FORCE TO SOMETHING BY SOMETHING ELSE IN DIRECT CONTACT WITH IT : COMPRESSION 3 *ARCHAIC* : IMPRESSION, STAMP 4 A : THE ACTION OF A FORCE AGAINST AN OPPOSING FORCE B : THE FORCE OR THRUST EXERTED OVER A SURFACE DIVIDED BY ITS AREA C : ELECTROMOTIVE FORCE 5 : THE STRESS OR URGENCY OF MATTERS DEMANDING ATTENTION : EXIGENCY

PRESSURE

IN PHYSICS, *PRESSURE* IS DEFINED as the magnitude of the force per unit of area. If we lean on a windowpane with an open hand, we apply a certain force that is distributed evenly over the area of the palm. The window probably will not break. But if we were to lean against the same window with a pointed object, such as the tip of a screwdriver (please don't try it!), the same force will be concentrated into a very small area and the window will probably break. The force applied is the same, but the *force per unit area* on the area of contact—the pressure—has increased.

All fluids—gas and liquid substances—are comprised of molecules that bounce around each other. When a fluid gets trapped inside a container, the constant collisions of particles against the container's walls result in a certain pressure. Consider a single molecule of a gas. The molecule hits the wall of the container with a certain velocity, then bounces off with a new velocity that has the same magnitude but a different direction. Since the velocity has changed, the molecule has experienced acceleration. According to the second law of dynamics, this acceleration should correspond to a force. According to the *third* law of dynamics, if the wall exerts a force on the molecule, it is equally true that the molecule exerts a force on the wall. If we sum the contributions of all the molecules hitting the wall, we get the total force on the wall. If we divide the total force by the area of the wall, we get the pressure.

Since each molecule contributes to the total force, the more particles there are, the denser the gas, and the higher the pressure. Further, the higher the velocity of the molecule, the higher the acceleration when it changes direction as it collides with the wall and the higher the force exerted on the wall by a single molecule. The temperature of the gas is a measure of the average speed of its particles. If the volume of the container does not change, a higher temperature corresponds to a higher pressure of the gas. We can increase the pressure of a bicycle tire by pumping more air into it, thereby increasing the density of the gas, but we could also warm up the tire by leaving the bike in the hot sun—as the air inside gets hotter, the pressure will increase.

See*: *Acceleration, Force, Heat, Velocity

$$p = \frac{F}{A}$$

[Eq. 9] ***The pressure (p) exerted on a surface of area A is equal to the force (F) perpendicular to the surface applied to it, divided by the area.***

THINK ABOUT IT: *Feeling the Pressure*

The pressure of the air at sea level is about 15 pounds per square inch, and it results from the weight of the atmosphere above us. This means that a 1-inch by 1-inch area of human skin experiences a force of 15 pounds due to *atmospheric pressure*. The reason we don't implode is that the pressure inside the body compensates for pressure outside of it, though the body does not easily adjust to rapid or extreme pressure changes. This is why we take measures to protect ourselves when we dive into deep water—in order to allow the body to adapt, we change our depth slowly. Similarly, airplane cabins are *pressurized* to protect passengers, as it would be impossible for us to breathe at such high altitudes where the pressure is so small.

READ ABOUT IT: *Environmental Advocacy & The Sea*

The Little Book of the Big Bang: A Cosmic Primer [1998] by Craig Hogan — Hogan's book is an accessible treatment of the theory of the history of the universe, complete with photographs and diagrams.

Silent Spring [1962] by Rachel Carson — This important text has been alerting readers to the dangers of chemical proliferation for decades. In 1999, *Time* magazine listed Carson as one of "the 20 most influential scientists and thinkers of the century." This fearless advocate of the environment (and of humanity!) has inspired generations of committed activists. Now available in a 40th-anniversary edition.

The Silent World [1953] by Jacques-Yves Cousteau with Frédéric Dumas — The memoir of the daring sea captain and his exploits on the great ship *Calypso* has been translated into 20 languages and has sold over 5 million copies. Cousteau's story of underwater exploration and the creation and use of the aqua-lung is thrilling. The film *Le Monde du Silence* by director Louis Malle (a friend of Cousteau's), won a Palme d'Or [1956], an Academy Award [1957], and a BAFTA [1957]. The book is now available in a 50th-anniversary edition.

TALK ABOUT IT: *Beyond Physics—Pressure & Landscape*

Consider pressure as the action of a force against an opposing force ... What forces have humans exerted over the force of nature—specifically, what pressures on the Earth have resulted from human and industrial over-development? ... How far beyond meeting the needs of shelter and community have we gone? ... At what natural cost have humans forced the planet to meet our demands? ... Can nature ever rebound from the burdens placed on it? ...

***The Great Wave at Kanagawa* [circa 1830-1832]**
by Katsushika Hokusai
from *A Series of Thirty-Six Views of Mount Fuji*
Polychrome woodblock print; ink on paper
Tokyo Fuji Art Museum, Tokyo

Hokusai spent eight years (between the ages of 64-72) making his *Thirty-Six Views of Mount Fuji,* an extremely important set of Japanese landscape pictures. *The Great Wave at Kanagawa* is said to have influenced composer Claude Debussy's *La Mer* and poet Rainer Maria Rilke's *Der Berg*. With the ocean wave curling hugely to engulf both the tiny distant peak of Mount Fuji and the pest-like fishing boats, the artist creates a visual *pressure*, demonstrating the volatile might of the sea. Hokusai became enamored with printmaking while working in a bookshop as a boy, and he then studied under a master woodblock artisan.

***The City in History* [1961] by Lewis Mumford**

Mumford traces the development of the city from ancient times to modern, drawing attention to the "power-trapping" capacity of the metropolis, and to the *pressures*—that is, the potentially negative consequences—that could result from the inorganic accumulation and concentration of resources. Writing in 1961, Mumford speaks prophetically of an "invisible city" a placeless system of creating and distributing goods that would link people from remote places. "There would be no limits, physical, cultural, or political, to such a system of co-operation," Mumford states. "It would pass through geographic obstacles and national barriers as readily as X-rays pass through solid objects ... Such a system in time could embrace the whole planet."[15] This description sounds amazingly like the Internet!

RADIATION 27

I waited. The sun was starting to burn my cheeks, and I could feel drops of sweat gathering in my eyebrows. The sun was the same as it had been the day I'd buried Maman, and like then, my forehead especially was hurting me, all the veins in it throbbing under the skin. It was this burning, which I couldn't stand anymore, that made me move forward. I knew that it was stupid, that I wouldn't get the sun off me by stepping forward. But I took a step, one step, forward. And this time, without getting up, the Arab drew his knife and held it up to me in the sun. The light shot off the steel and it was like a long flashing blade cutting at my forehead. At the same instant the sweat in my eyebrows dripped down over my eyelids all at once and covered them with a warm, thick film. My eyes were blinded behind the curtain of tears and salt. All I could feel were the cymbals of sunlight crashing on my forehead and, indistinctly, the dazzling spear flying up from the knife in front of me. The scorching blade slashed at my eyelashes and stabbed at my stinging eyes. That's when everything began to reel. The sea carried up a thick, fiery breath. It seemed to me as if the sky split open from one end to the other to rain down fire. My whole being tensed and I squeezed my hand around the revolver. The trigger gave; I felt the smooth underside of the butt; and there, in that noise, sharp and deafening at the same time, is where it all started. I shook off the sweat and sun.

L'ÉTRANGER [THE STRANGER]
ALBERT CAMUS
TRANSLATED BY MATTHEW WARD

From: **RA-DI-A-TION** *n* (15C)

1 A : THE ACTION OR PROCESS OF RADIATING B : THE PROCESS OF EMITTING RADIANT ENERGY IN THE FORM OF WAVES OR PARTICLES C (1) : THE COMBINED PROCESSES OF EMISSION, TRANSMISSION, AND ABSORPTION OF RADIANT ENERGY (2) : THE TRANSFER OF HEAT BY RADIATION—COMPARE CONDUCTION, CONVECTION 2 A : SOMETHING THAT IS RADIATED B : ENERGY RADIATED IN THE FORM OF WAVES OR PARTICLES

RADIATION

IN PHYSICS, THE WORD *RADIATION* is used whenever energy propagates away from its source along straight lines. The word derives from the Latin "radius," in that radiation occurs when energy moves away from a source "radially," as the radius of a circle "moves" away from the center. Energy can be radiated in the form of particles or waves.

The radiation we are most familiar with is light. We know that light propagates along a straight line because we observe shadows: light does not turn around objects. Light is electromagnetic radiation, which is described in classical physics as a wave with frequency in a particular range of values, the *visible range*, though there are electromagnetic waves of very different frequencies, most of which cannot be perceived by the human eye. A bulb emits light—*visible electromagnetic radiation*—but every object, by virtue of being at a certain temperature, no matter how small, is a source of electromagnetic radiation. Our bodies emit electromagnetic radiation in the *infrared* range of frequencies. It is called infrared because the range starts just below the frequency of red light, and red is the smallest frequency we can perceive. Infrared goggles do not allow us to "see" in darkness, but to detect infrared radiation emitted by warm objects in the absence of light. Though we cannot perceive infrared radiation without the help of equipment, we can feel it on our skin. While sitting in front of a fire, we can feel that the parts of our body exposed to the firelight are warmer than the ones in shadow. This simple observation shows that most of the heat we feel comes from the source in a straight line, as does light. What warms our skin is not light, though, but infrared radiation.

In other forms of radiation, energy is transported by particles. Radioactive materials emit radiation through nuclear decay. As the unstable nucleus of an atom decays, it gives off *alpha*, *beta*, and *gamma* rays. Alpha rays are composed of helium nuclei, a bundle of two protons and two neutrons. Beta rays are electrons or positrons, the antiparticles of electrons, moving at very high speed. Gamma rays are electromagnetic radiation of very high frequency, far above the visible spectrum.

***See:* Energy, Heat, Light, Particle, Wave**

THINK ABOUT IT: ***Radioactivity***

Radioactivity is the name given to the spontaneous emission of *alpha*, *beta*, and *gamma* rays by some elements. The occurrence was so named because at the time of its discovery, although it was not clear what exactly was emitted, it was clear that it must have been some sort of radiation. The discovery of radioactivity was the first observed manifestation of two completely new fundamental forces: the *strong nuclear force*, responsible for keeping the nucleus of the atom together, and the *weak nuclear force*, responsible for the emission of beta rays. Together with the *electromagnetic force* and the *force of gravity*, they make up the four basic forces we need to describe the observed physical world.

READ ABOUT IT: ***Nuclear Disasters—Real & Imagined***

On the Beach [1957] by Nevil Shute — This stirring story of the last months on Earth following a nuclear disaster inspired legions of activists, including Helen Caldicott of the Nuclear Policy Research Institute. The movie [1959] stars Gregory Peck and Ava Gardner.

Hiroshima-Nagasaki, August 1945 [1970] produced by Erik Barnouw — The Japanese government had a Japanese filmmaker take footage of the aftermath of the atomic bombs the U.S. dropped on the cities of Hiroshima and Nagasaki. In 1970, Barnouw, then chair of Columbia University's media department, learned that the footage was being stored in Washington, D.C., and he asked for permission to use it. Upon seeing the footage—edited by Barnouw—for the first time in 25 years, the original director, Akira Iwasaki, was moved. He wrote, "This film is an appeal or warning from man to man for peaceful reflection—to prevent the use of the bomb again."[16]

No Nukes [1979] directed by Daniel Goldberg — Six months after the meltdown of the nuclear reactor at Three Mile Island in Pennsylvania, Bruce Springsteen, Carly Simon, James Taylor, Bonnie Raitt, Jackson Browne, and others performed to raise awareness and to protest.

TALK ABOUT IT: ***Beyond Physics—Radiation & Joy***

If radiation can be thought of as energy moving away from a source, what sorts of energies do people radiate? ... Consider the transmission of hostility, serenity, joy ... How do people manage their emotional states in public? ... What might be the benefits of seclusion, modesty, celebration? ... How do rituals (graduations, weddings, funerals) provide people with the opportunity to share energies? ...

***Evening Star No. VI* [1917] by Georgia O'Keeffe**
Watercolor on paper
Georgia O'Keeffe Museum, Santa Fe

Wisconsin-born O'Keeffe knew from an early age that she wanted to be an artist. Though perhaps best known for her renderings of skulls and flowers, she also created works like this *radiant* watercolor, which is as abstract, symbolic, and triumphant as a flag. She was married to gallery owner and art photographer Alfred Stieglitz (See: *Angular Velocity*), and they settled—somewhat famously—in Sante Fe. It comes as no surprise that the artist would be attracted to a place as iconically American as New Mexico since there is a uniquely American quality to her imagery. She is equally at home with the vast and the intimate, with the public and the private. With O'Keeffe's work, one is conscious of her having rendered a thing *completely*: as both grand, communal icon and delicate, personal gesture.

***L'Étranger [The Stranger]* [1942] by Albert Camus**
Translated by Matthew Ward

In the most unforgettable scene in this most unforgettable book, the typically impassive Meursault is brought to carnal life by the blinding light of the sun on an Algiers beach. The jarring effects of the *radiation* of the sun compromise his reason and his consciousness, and, in a meaningless act of free will, he kills a man. Ward's translation of this chilling story captures the flat and detached tone of the author's "voice," which, in turn, emphasizes the willfully existential essence of the tightly wound narrative. More than a study of character, *The Stranger* is a study of the philosophy of being.

REFLECTION 28

Without therapy, it is impossible for the grandiose person to cut the tragic link between admiration and love. He seeks insatiably for admiration, of which he never gets enough because admiration is not the same thing as love. It is only a substitute gratification of the primary needs for respect, understanding, and being taken seriously—needs that have remained unconscious since early childhood. Often a whole life is devoted to this substitute.... The grandiose person is never really free; first, because he is excessively dependent on admiration from others, and second, because his self-respect is dependent on qualities, functions, and achievements that can suddenly fail.

THE DRAMA OF THE GIFTED CHILD
ALICE MILLER

We have seen that the music which a listener hears in a reverberant room consists of different parts; one comes direct from the musical instrument, but other parts, almost equal in intensity, reach him after one, two, three or more reflections. The majority of the sound which enters his ear has been reflected many times.... Sabine's theory, at least in its simplest form, proceeds on the supposition that the energy of sound is spread uniformly through the hall, so that every cubic foot of space contains the same amount. This is of course accurately true only in a very reverberant room. The energy does not stand still, but travels in all directions at a speed of 1100 feet a second, and this results in its extinction. A stream of sound energy is continually falling on all the walls and pieces of furniture in the room. If these were perfectly reflecting, no energy could be absorbed, and the sound would stay always at the same level of intensity.

SCIENCE & MUSIC
SIR JAMES JEANS

From: **RE-FLEC-TION** *n* (14C)
1 : AN INSTANCE OF REFLECTING; *ESPECIALLY* : THE RETURN OF LIGHT OR SOUND WAVES FROM A SURFACE 2 : THE PRODUCTION OF AN IMAGE BY OR AS IF BY A MIRROR 3 A : THE ACTION OF BENDING OR FOLDING BACK

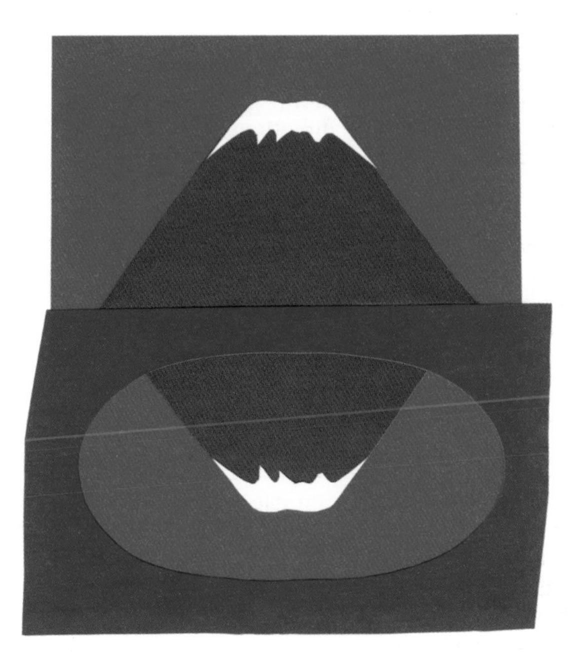

REFLECTION

A *REFLECTION* OCCURS ANY TIME a wave traveling through some *medium* meets the *interface* with another medium. We hear an echo, for example, when a sound wave traveling in air—the medium—meets the surface of a mountain or a building—the interface—and bounces back. If we drop a stone in the center of a swimming pool, we will see waves propagating radially from the center to the edge, and we will see waves bouncing back at the edge, where they reflect off the wall.

We see objects—shapes and colors—because they reflect light, which is an electromagnetic wave. When we think of the reflection of light, we usually think of a mirror, but, in fact, everything reflects light. The main difference between the way a mirror reflects and the way a wall reflects is that the wall has a more irregular surface and reflects parallel beams of light in many directions. In the case of a mirror, the reflected beam makes an angle with the surface exactly equal to the angle made by the incoming beam. If the incoming beams are parallel to each other, the outgoing beams are parallel as well and result in a sharp image. This phenomenon can be understood by considering a mountain reflected in a lake—if the water is perfectly still, we can see the reflected image more clearly than if the water is moving. A mirror reflects a beam of light in roughly the same way a ball bounces off the ground. This is why light was once thought to be comprised of particles traveling straight from the light source, bouncing off an object, and reaching our eyes.

Geometrical optics, the geometrical description of reflection and refraction of light, is relatively simple and was known long before the nature of light as an electromagnetic waves was understood in the middle of the 19th century. In order to build lenses and telescopes, it was enough to know that light travels in straight lines and to know how it changes direction when it meets another material such as glass or water. According to legend, the Ancient Greek mathematician and inventor, *Archimedes* [287 B.C.E.-212 B.C.E.] built bronze mirrors that could reflect sunlight, and he concentrated the rays on the sails of enemy Roman ships in order to set the sails on fire before the ships could enter the harbor of Syracuse.

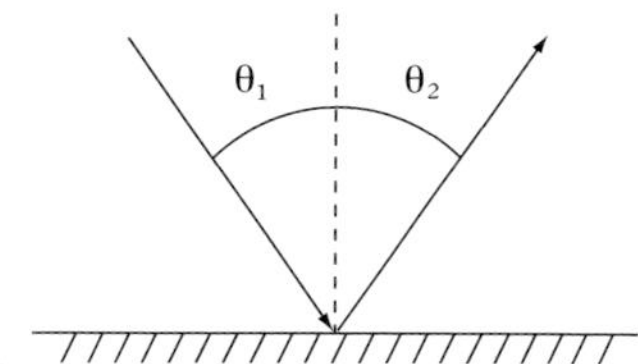

See: ***Angle, Color, Image, Light, Refraction, Wave***

[Ill. 3] ***The angle of incidence (θ_1) is equal to the angle of reflection (θ_2). The horizontal line represents the surface, while the dashed line represents the line normal (perpendicular) to the surface. Angles of incidence and reflection are measured with respect to the normal line.***

THINK ABOUT IT: *Fiber Optics*

When a beam of light meets a surface, part of it is *reflected* while the rest is usually *refracted* and enters the surface. In some cases, if the angle the beam makes with the surface is very small, the beam is totally reflected—there is no refraction. A fiber optics cable is a small plastic tube with a reflecting inner surface that makes use of total internal reflection. Its purpose is to carry light, and, by means of light signals, information. If the beam of light is always totally reflected, it does not get lost along the fiber. Information can be moved via *fiber optics* without essential energy loss, whereas moving information via electric signals requires energy to move electric charges along the cable.

READ ABOUT IT: *Thinking Critically About Art & Light*

Illuminations: Essays and Reflections [1955] by Walter Benjamin, edited by Hannah Arendt — The essays of this important critical thinker are thoughtful and accessible. Benjamin wrote provocatively about art—literature, theater, film, etc.—and about those who made art—Kafka, Baudelaire, Proust, et al. In the legendary and still relevant "The Work of Art in the Age of Mechanical Reproduction" [1936], Benjamin discusses modes of artistic reproduction for the masses, touching on issues of authenticity, remembrance, reflection, separation, transportation, and perception of the image.

Introduction to Light: The Physics of Light, Vision, and Color [1983] by Gary Waldman — This nicely illustrated discussion of light has an "old school" science appeal. There are chapters on optical illusions, holograms, rods and cones, waves and polarization, and theories of color. This is a fine resource with especially clear diagrams.

TALK ABOUT IT: *Beyond Physics—Reflection & Art*

In art, as in science, the concept of reflection describes a "bouncing" or a "bending" back ... In reflexive art, authors or artists might incorporate themselves, their materials, or the circumstances of production into the art itself ... A film actor might directly address the camera, or a director might appear in a scene (e.g., Hitchcock) ... The self-conscious break from traditional methods offers a commentary on realism versus artifice (i.e., art that features its own production announces that it is "not real") ... Reflexive art invites self-evaluation, self-scrutiny, and intellectual reflection: it encourages people to see themselves as subjects and objects ... Consider some contemporary artists who refuse to engage in the illusion of continuity, preferring instead to make art that provokes audiences ...

***Las Meninas (The Maids of Honor)* [1656]**
by Diego Velázquez
Oil on canvas
Museo del Prado, Madrid

In this sensational painting, Infanta Margarita is surrounded by her Maids of Honor, or *Meninas*, as they were called in 17th-century Portuguese. In addition to the *reflection* of the artist, who painted himself painting a canvas, a mirror reflects the figures of Queen Mariana and King Philip IV. *Las Meninas* has been studied by painters, mathematicians, philosophers, poets, and astrologists. Through the use of direct and indirect light, Velázquez creates the impression of several stories occurring at once. According to *The Science of Art*, which contains a perspectival analysis of the work, "no painting has ever been more concerned with looking."[17]

***The Drama of the Gifted Child* [1979] by Alice Miller**

According to Miller, few escape the early years without experiencing some trauma, humiliation, or loss. Believing that rational behaviors are born from access to true feelings, Miller suggests that the *reflection* done during the therapeutic process is the best means of coping.

***Science & Music* [1937] by Sir James Jeans**

The respected British scientist describes the mechanics of sound—how notes are created and *reflected*. Jeans explains the science of perfect performance acoustics and the way music moves through the ears to be interpreted by the brain.

REFRACTION 29

SECOND VOICE

laughs high and aloud in his sleep and curls up
his toes as he sees, upon waking fifty years ago,
snow lie deep on the goosefield behind the sleeping house;
and he runs out into the field where his mother is making
welshcakes in the snow, and steals a fistful of snowflakes
and currants and climbs back to bed to eat them cold and
sweet under the warm, white clothes while his mother
dances in the snow kitchen crying out for her lost currants.

FIRST VOICE

And in the little pink-eyed cottage
next to the undertaker's, lie, alone, the seventeen
snoring gentle stone of Mister Waldo, rabbitcatcher, barber,
herbalist, catdoctor, quack, his fat pink hands, palms up,
over the edge of the patchwork quilt, his black boots neat
and tidy in the washing-basin, his bowler on a nail above
the bed, a milk stout and a slice of cold bread pudding
under the pillow; and, dripping in the dark, he dreams of

UNDER MILK WOOD: A PLAY FOR VOICES
DYLAN THOMAS

From: **RE-FRAC-TION** *n* (1603)
1 : DEFLECTION FROM A STRAIGHT PATH UNDERGONE BY A LIGHT RAY OR ENERGY WAVE IN PASSING OBLIQUELY FROM ONE MEDIUM (AS AIR) INTO ANOTHER (AS GLASS) IN WHICH ITS VELOCITY IS DIFFERENT 2 : THE CHANGE IN THE APPARENT POSITION OF A CELESTIAL BODY DUE TO BENDING OF THE LIGHT RAYS EMANATING FROM IT AS THEY PASS THROUGH THE ATMOSPHERE; *ALSO* : THE CORRECTION TO BE APPLIED TO THE APPARENT POSITION OF A BODY BECAUSE OF THIS BENDING 3 : THE ACTION OF DISTORTING AN IMAGE BY VIEWING THROUGH A MEDIUM

REFRACTION

LIKE REFLECTION, *REFRACTION* OCCURS WHENEVER a wave traveling through some medium meets the interface with another medium. Refraction refers in particular to the way in which a wave changes its direction of propagation as it enters a different substance. A common example of refraction is that of a straw standing in a glass of water, where the straw appears to be bent. As the ray of light crosses the surface from water to air, the ray is bent, and the direction of the ray as it reaches the eye causes the straw to appear to be standing where actually it is not.

Refraction is the principle behind the functioning of every sort of lens. Due to its curved surface, a lens exploits the process of refraction by causing rays of light to converge or diverge as they enter the glass from the air, and as they leave the glass to the air. Because it is so easy to observe, refraction has been known since antiquity—though the first glasses to make use of the properties of refraction to correct sight defects did not appear until the Middle Ages. The mathematical description of the geometry of refraction found its final form in 1621 with the Dutch mathematician *Willebrord Snellius* [1591-1626] and Snell's Law. An explanation that came shortly after Snell's Law for the bending of the ray of light was given by French mathematician *Pierre de Fermat* [1601-1665]. Assuming that the speed of light is different in different materials, Fermat's Principle [1658] states that light "chooses" the path that minimizes the time it takes to go from one point to another. Such a path doesn't necessarily have to correspond to the shortest distance. Let's return to the example of the straw in the glass. The ray of light is actually bent at the surface of the water in order to travel a *shorter path* in the water with respect to what it would travel underwater if it were going straight from the straw to our eyes. This might sound complicated, but, actually, it can be understood intuitively. To save a swimmer, an ocean lifeguard would run along the beach almost to a point in front of the drowning person and then swim the *shortest distance possible* in order to arrive in the fastest time. The overall distance would be longer than if the lifeguard had gone in a straight line from his initial position to the swimmer, but the time would in fact be shorter.

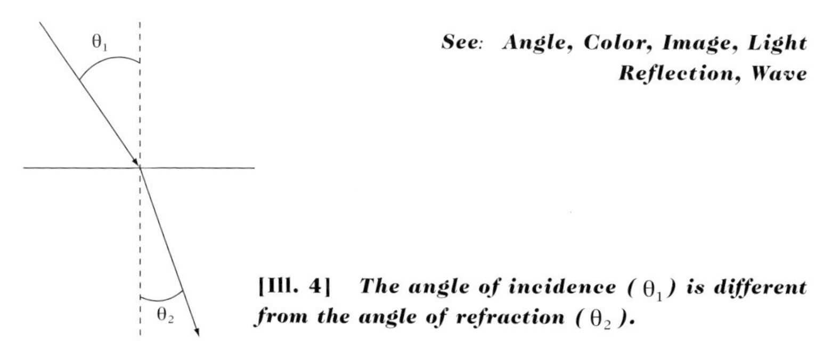

***See:* Angle, Color, Image, Light Reflection, Wave**

[Ill. 4] ***The angle of incidence (θ_1) is different from the angle of refraction (θ_2).***

THINK ABOUT IT: ***Rainbows***

Rainbows can be explained in terms of *reflection* and *refraction* of light rays passing through rain drops. A light ray hitting a raindrop is first refracted, then it reflects inside the drop, and finally it refracts out of the drop to reach our eyes. Since every color corresponds to a different *frequency*, and the angle of refraction depends on the frequency of the light, rays of light of different colors leave the drop in different directions, pretty much as happens in a prism. Sometimes we can see two rainbows at the same time. The less intense one comes from light rays that have been reflected two times inside the drop before being refracted out.

READ ABOUT IT: ***Northern Lights, Making Holograms at Home, & Lighting the Image***

Aurora: The Northern Lights in Mythology, History and Science [1983] by Harald Falck-Ytter — In their efforts to describe the beauty of auroras, people have often turned to art. The author uses samples of medieval woodcuts, German engravings, Danish lithographs, and modern art, as well as poetry, history, and mythology to describe the mysterious phenomenon of the northern lights.

The Complete Book of Holograms [1987] by Joseph Kasper and Steven Feller — Though the physics of holography involves the use of abstract expressions to represent light waves and mathematics to explore interference, the authors give clear instructions for making a hologram, including material requirements such as inner tubes for building a table that will not move "one-half a wavelength of light."

Painting with Light [1949] by John Alton — Alton's book demonstrates the scientific basis of the use of light in photography. Written by an Academy Award-winner (*An American in Paris* [1951]), it provides a behind-the-scenes look at Hollywood filmmaking.

TALK ABOUT IT: ***Beyond Physics—Refraction & Art***

In physics, refraction refers to a bending, a deflecting, a distorting ... In art, Surrealism confronts the fact that although we live in a world of logic, there are aspects of human experience that logic cannot address ... How is refraction similar to Surrealism? ... Surrealism celebrates the refracted realms of the illogical, the nonlinear, and the irrational (e.g., dreams, the subconscious), as well as chance, surprise, and randomness ... It aims to challenge familiar expectations and tap into deeper, purer levels of being, sensation, knowledge ...

***Blocks and Strips* [circa 1970] by Annie Mae Young**
Cotton, polyester, synthetic blends
Collection of the Tinwood Alliance, Atlanta

The African-American women of Gee's Bend, Alabama, have been making quilts of distinction for six generations. These profoundly beautiful objects tell a complicated story of modern American design, cultural transcendence, and female identity. The blankets, often made from old work clothes, received the newly-born, warmed the cold, cradled the elderly, and covered the dead. It seems that the light that shines on them should change direction under our gaze, *refracting* as we "interface" with a people through the medium of fabric.

***Under Milk Wood: A Play for Voices* [1952]**
by Dylan Thomas

"It is Spring, moonless night in the small town, starless and bible-black." So begins the sublime play Thomas worked on for ten years before completing it one month prior to his death at age 39. This iconoclastic Welsh master poet is remembered for having deviated from the straight path both in language *and* in life. His choices are indicative of artistry at its finest—personal, original, and free—and his style is deceptively simple. Familiar words *refract* as they pass through the medium of the author's mind, turning oblique, giving one the sense of encountering something entirely new: e.g., *rabbit-catcher*, *catdoctor*, *snow kitchen*, *pink-eyed cottage*, and, *dripping in the dark he dreams of—*

RELATIVITY 30

Cubism is no different from any other school of painting. The same principles and the same elements are common to all. The fact that for a long time Cubism has not been understood and that even today there are people who cannot see anything in it, means nothing. I do not read English, an English book is a blank book to me. This does not mean that the English language does not exist, and why should I blame anybody else but myself if I cannot understand what I know nothing about?

PICASSO SPEAKS, 1923
PABLO PICASSO
FROM: ART IN THEORY: 1900-1990:
AN ANTHOLOGY OF CHANGING IDEAS
EDITED BY CHARLES HARRISON & PAUL WOOD

From: **REL-A-TIV-I-TY** *n* (CIRCA 1834)
1 A : THE QUALITY OR STATE OF BEING RELATIVE B : SOMETHING THAT IS RELATIVE 2 : THE STATE OF BEING DEPENDENT FOR EXISTENCE ON OR DETERMINED IN NATURE, VALUE, OR QUALITY BY RELATION TO SOMETHING ELSE 3 A : A THEORY WHICH IS BASED ON THE TWO POSTULATES (1) THAT THE SPEED OF LIGHT IN A VACUUM IS CONSTANT AND INDEPENDENT OF THE SOURCE OR OBSERVER AND (2) THAT THE MATHEMATICAL FORMS OF THE LAWS OF PHYSICS ARE INVARIANT IN ALL INERTIAL SYSTEMS AND WHICH LEADS TO THE ASSERTION OF THE EQUIVALENCE OF MASS AND ENERGY AND OF CHANGE IN MASS, DIMENSION, AND TIME WITH INCREASED VELOCITY

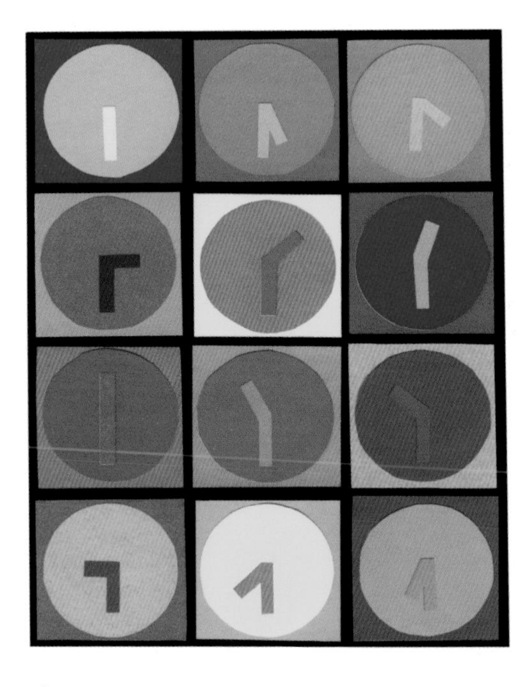

RELATIVITY

IMAGINE WE COULD TRAVEL ON a train that has no windows and that can move without vibrations. As the train leaves the station, we would know that it is accelerating because we would feel pushed back into our seat. But if we were to fall asleep on this strange train and then wake up later, we would not be able to discern if the train was standing still or moving at a constant velocity, with a constant speed and direction. If, in fact, the velocity of the train was constant, we could get up and walk around or even juggle a ball as well as we would be able to do on the platform of the station. We would see a ball fall on the floor of the train just as it would fall on the floor of the station.

Observing objects moving around us, we would assume the train to be our *frame of reference*. But if the train were to make a curve, walking and juggling would not be as easy as they previously were because we would feel *accelerated toward the side of the train*. The same holds true if the train slows down—we would feel a *forward acceleration*. According to the second law of dynamics, whenever we observe an acceleration, some force must be responsible for it. Since we are on a train that is not moving at constant velocity, we are measuring accelerations that do not seem to correspond to any actual force, but are due to the fact that the train itself is accelerated. In physics we say that we are in a *non-inertial frame of reference*, that is, a frame of reference in which the principle of inertia does not hold. In a non-inertial frame, it is *not true* that if no force is acting on a body it would keep its velocity constant. As the train turns, we would see a ball on a table accelerating toward the side of the train.

We can assume the station to be an *inertial frame of reference*, the frame of reference where any acceleration corresponds to an actual force. The same would be true also for a train, or any other frame of reference, moving at constant velocity relative to the station. In general, we can say that the laws of dynamics—the laws studying force and acceleration—are the same for two observers moving relative to each other at constant velocity, that is constant speed and direction. This is called the *principle of Galilean relativity*, since Galileo was the first to explicitly state this principle.

See:* *Acceleration, Equation, Gravity, Measurement, Space-time, Velocity

THINK ABOUT IT: *Special and General Relativity*

When Einstein tried to extend the Galilean principle of relativity to the laws describing electricity and magnetism, he realized it was possible only by assuming that time flows differently for the two inertial frames moving at constant velocity, one with respect to the other. The time it takes for the train to pass through a station at a certain speed is different if measured by the passenger on the train than if by someone on the platform. In practice, if the train is moving at ordinary speed, the difference would be too small to be measured, which is why it is natural for us to assume an *absolute time*, valid for every observer. By dropping the concept of absolute time, Einstein was able to state that the laws of dynamics, as well as those of electromagnetism, are the same for two observers moving relative to each other at constant velocity. This is called the *principle of special relativity*. Einstein later formulated a principle of relativity valid for all physical phenomena, including gravitation, which led to the *theory of general relativity*.

READ ABOUT IT: *Einstein & Mr. Tompkins*

Ideas and Opinions [1954] by Albert Einstein — In these immensely personal writings, Einstein discusses religion, education, individual conscience, human rights, art, pacifism, and the work of other scientists. There is an obituary for Marie Curie, a letter to Sigmund Freud, a review of Bertrand Russell, a birthday note to Mahatma Gandhi, and a succinct explanation of relativity. Einstein demonstrates that being a scientist is not incompatible with expressing emotion and possessing a global view of humanity and interconnectedness.

Mr. Tompkins in Paperback [1965] by George Gamow — A funny look at what things would be like if certain basic physical constants were different. Tompkins, protagonist of this illustrated fiction, is a bank teller who dreams of different realities (e.g., where the speed of light is slow or where quantum behavior manifests on large scales).

TALK ABOUT IT: *Beyond Physics—Relativity & Looking*

Self/other ... What do we seek when we look? ... What is it that a view provides to the viewer? ... How is forming an opinion of everyone else the simplest way for us to constitute an idea of ourselves? ... How do we think about "other" societies? ... What do we hope to learn about them that is different from what we know about ourselves? ... Do we seek true understanding, or just the power to hold ourselves separate, to define ourselves by what we are not? ...

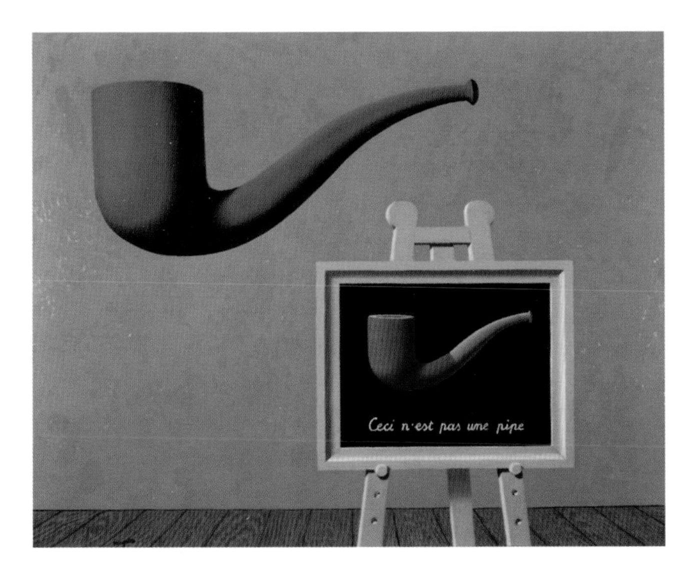

***Les Deux Mystéres (Two Mysteries)* [1966]**
by René Magritte
Oil on canvas
Private Collection

This is not a pipe—it is a painting of a pipe! Naturally, the *representation* of an object is not the *actual* object, but Magritte's witty distortion unseats the viewer regardless, leading to questions relating to perception, language, position, and, especially, the nature of reality. Like Einstein's theory of *relativity*, which maintains that it is not possible to look at a problem in isolation (since everything is dependent upon everything else), Magritte's surrealist perspective exposes the divide between what we think we know and see, and whether or not anything is in fact knowable or seeable.

"Picasso Speaks" [1923], from *Art in Theory: 1900-1990: An Anthology of Changing Ideas* [1992]
Edited by Charles Harrison and Paul Wood

Picasso's essay reveals the existential zeitgeist that marked the early 20th century in the wake of the social and intellectual upheavals caused by industrialization, modernization, and Einstein's theory of *relativity*, which, though difficult to understand, transformed the era. Newton's mechanical universe had been replaced by an unfathomable entity, something that is both infinitely whole and infinitely particulate. Regardless of artistic intention, Cubism as a mode of expression depicting the fracturing of time and space (See: *Space-time*) reflected changes in modes of perception.

SPACE-TIME

And an astronomer said, Master, what of Time?
And he answered:
You would measure time the measureless and the immeasurable. You would adjust your conduct and even direct the course of your spirit according to hours and seasons. Of time you would make a stream upon whose bank you would sit and watch its flowing.

Yet, the timeless in you is aware of life's timelessness, and knows that yesterday is but today's memory and tomorrow is today's dream. And that that which sings and contemplates in you is still dwelling within the bounds of that first moment which scattered the stars into space.
Who among you does not feel that his power to love is boundless? And yet who does not feel that very love, though boundless, encompassed within the centre of his being, and moving not from love thought to love thought, nor from love deeds to other love deeds? And is not time even as love is, undivided and spaceless?

But if in your thought you must measure time into seasons let each season encircle all the other seasons. And let today embrace the past with remembrance and the future with longing.

THE PROPHET
KAHLIL GIBRAN

From: **SPACE-TIME** *n* (1915)
1 : A SYSTEM OF ONE TEMPORAL AND THREE SPATIAL COORDINATES BY WHICH ANY PHYSICAL OBJECT OR EVENT CAN BE LOCATED—CALLED ALSO *SPACE-TIME CONTINUUM* 2 : THE WHOLE OR A PORTION OF PHYSICAL REALITY DETERMINABLE BY A USUALLY FOUR-DIMENSIONAL COORDINATE SYSTEM

SPACE-TIME

When we think about space, we tend to think about the objects we see—the places they occupy and the distances between them. When we think about time, we tend to think about moments and about the intervals between them—seconds, hours, days, years. It goes without saying that the length of a table—the distance between two points in space—is some absolute quantity that everyone can measure more or less accurately. It also goes without saying that the time interval between color changes in a traffic light is an absolute quantity.

But in the last century, it became clear to physicists that these are just our impressions. The way nature behaves, as described by the theory of special relativity, is different from what common sense would suggest. If we could move at a very high speed—close to the speed of light—with respect to the table, we would realize that we would measure a different value for its length than if we were moving at a "normal" speed. The same would be true if we were to measure the time intervals between red lights at an intersection. This does not mean that there is anything wrong with our measurements. It means that distances and time intervals depend upon the speed of the observer. They are relative, not absolute, quantities. The reason this fact is not obvious to us is that we cannot move at speeds close to the speed of light.

However, not everything is relative. Instead of thinking in terms of *positions in space* and *moments in time*, we can think of events in *space-time*. An event is a certain place at a certain time. If one end of the table is a certain point in space, that same end considered at a certain moment in time is an event in *space-time*—a point in a four-dimensional space. It is possible to define a distance between events in space-time given by a particular combination of space intervals and time intervals. Two observers moving at a certain velocity, one with respect to the other, will not agree on distances in space or intervals in time, but they will agree on the "distance" between two events in space-time. Space-time allows us to define quantities independent of the state of motion of the observer. The theory of special relativity forces us to drop the intuitive notion of space and time as absolutes.

See:* *Light, Measurement, Position, Relativity

THINK ABOUT IT: *The Speed of Light*

One of the postulates of the *theory of special relativity* is the fact that the speed of light is constant for any observer, in any *frame of reference*. When we throw a ball inside a train, in the train's direction of motion, we can measure the ball's speed with respect to the train and find a certain value, say 30 miles per hour. If someone on the ground measures the speed of the train to be 80 mph, they would measure the speed of the ball in the train to be 110 mph. But if we shine a beam of light on the train, the speed of light we would measure on the train is exactly the same as the speed of light that would be measured on the ground, no matter how fast the train goes. This was proven to be true in experiments (not done on trains) at the end of the 19th century.

READ ABOUT IT: *Diary of a Physicist & Field Theory*

How the Universe Got Its Spots [2002] by Janna Levin — This book takes an engaging approach to physics: it is written in diary form. Levin, whose career includes time at MIT, Cambridge University, the Canadian Institute for Theoretical Astrophysics, and the Center for Particle Astrophysics at the University California, Berkeley, weaves the history of physics with contemporary data, trends, and studies.

Hyperspace [1994] by Michio Kaku — This is a clear and thorough general science book for those curious about the possible beginnings of the universe and the "mystery" of field theory. Kaku is a professor of theoretical physics at City College of New York and the host of *Explorations in Science* on Pacifica Radio.

Groundhog Day [1993] directed by Harold Ramis — This glorification of the fourth dimension set to a Sonny and Cher song suggests that answers to meaningful living can be found in transcending the limits of space-time—or in keeping the mind elastic and the heart pure.

TALK ABOUT IT: *Beyond Physics—Space-time & The Idea of Coincidence*

The concept of space-time seems to refer to a multiplicity of coinciding perspectives ... What is the importance of coincidence to the psychic life of humans? ... Why does the sense of being in the right place at the right time reassure and validate us? ... How does the depiction of random but nonetheless meaningful events in the television show Seinfeld *allude to the principles of space-time—i.e., "everything depends on the perspective of the observer" and "the world is different from what common sense would suggest"? ...*

***Les Demoiselles d'Avignon* [1907] by Pablo Picasso**
Oil on canvas
The Museum of Modern Art, New York

Every artwork treats the illusion of space differently. Prior to the Renaissance, artists used vertical or overlapping perspective; during the Renaissance, they tested "scientific," linear perspective; and, eventually, near the beginning of the 20th century, they experimented with distorted (elongated or collapsed) perspective. In Cubism, there is a multiplicity of perspectives, a denial of spatial or textual integrity, a revolt against what can be known, and an aesthetic observance of the mystery of *space-time*. Looking into this painting is like looking into a web of shattered glass, into a field of events at once related and unrelated. The women swell forward asymmetrically, intruding brazenly upon the viewer's space. Incredibly, this triumph of Cubism, so revolutionary and ahead of its time, was kept facing a wall in Picasso's studio for 30 years after first being shown.

***The Prophet* [1923] by Kahlil Gibran —**

This poetic examination of the quest for meaning has been reprinted 135 times and translated into 20 languages. Nine million copies have sold in the U.S. alone. Clearly, the author touched on something universally meaningful when he wrote this philosophy of active existence. In brief conversational segments, *The Prophet* of the title gently and generously addresses questions concerning sorrow, joy, reason, love, anger, prayer, and—it would seem—*space-time*.

SPIN 32

É pau, é pedra,	*A stick, a stone,*
é o fim do caminho	*it's the end of the road*
É um resto de toco,	*It's the rest of a stump,*
é um pouco sozinho	*it's a little alone*
É um caco de vidro,	*It's a sliver of glass,*
é a vida, é o sol	*it is life, it's the sun,*
É a noite, é a morte,	*It is night, it is death,*
é o laço, é o anzol	*it's a trap, it's a gun*
É peroba do campo,	*The oak when it blooms,*
é o nó da madeira	*a fox in the brush*
Caingá, candeia,	*A knot in the wood,*
é o matita pereira	*the song of a thrush*
É madeira de vento,	*The wood of the wind,*
tombo da ribanceira	*a cliff, a fall*
É o mistério profundo,	*A scratch, a lump*
é o queira ou não queira	*it is nothing at all*
É o vento ventando,	*It's the wind blowing free,*
é o fim da ladeira	*it's the end of the slope*
É a viga, é o vão,	*It's a beam, it's a void,*
festa da cumieira	*it's a hunch, it's a hope*
É a chuva chuvendo,	*And the river bank talks*
é conversa ribeira	*of the waters of March*
Das águas de março,	*It's the end of the strain*
é o fim da canseira	*The joy in your heart*

ÁGUAS DE MARÇO [WATERS OF MARCH]
ANTONIO CARLOS JOBIM

From: **SPIN** *n* (1831)

3 A : A QUANTUM CHARACTERISTIC OF AN ELEMENTARY PARTICLE THAT IS VISUALIZED AS THE ROTATION OF THE PARTICLE ON ITS AXIS AND THAT IS RESPONSIBLE FOR MEASURABLE ANGULAR MOMENTUM AND MAGNETIC MOMENT B : THE ANGULAR MOMENTUM ASSOCIATED WITH SUCH ROTATION WHOSE MAGNITUDE IS QUANTIZED AND WHICH MAY ASSUME EITHER OF TWO POSSIBLE DIRECTIONS; *ALSO* : THE ANGULAR MOMENTUM OF A SYSTEM OF SUCH PARTICLES DERIVED FROM THE SPINS AND ORBITAL MOTIONS OF THE PARTICLES

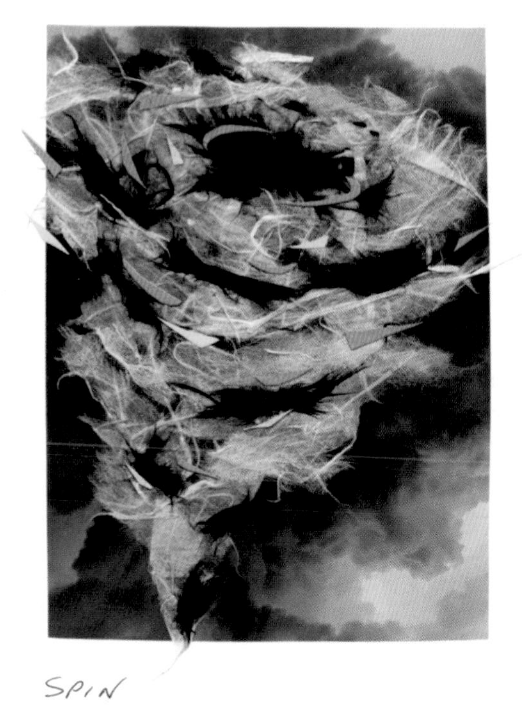

SPIN

WHENEVER AN ELECTRIC CHARGE ROTATES around an axis, it will interact not only with an external electric field but also with a magnetic field, just as the needle of a compass does. In particular, it will have the tendency to align its axis of rotation with the magnetic field. Physicists assign a *magnetic moment* to the needle or to the rotating charge to measure how sensitive they are to a given magnetic field. A magnetic moment can also be assigned to a charge with a certain size rotating on itself. Although the electron is an elementary particle—something considered to be "size-less"—physicists nevertheless assign to the electron an intrinsic magnetic moment called *spin*. This does not mean that electrons have a size or that they are, in fact, *spinning*. We have no evidence of that.

Scientists first studied the properties of electrons by exciting the electrons in atoms. An electron is excited when its energy increases. The electrons that make up an atom cannot have any energy they "want." They are bound to well defined energy levels. Such levels are *discrete*—they can be numbered. By shining light on atoms, we can make the electron jump from one level to another level of higher energy. The electron can then return to the original level by emitting a *photon*, a particle of light. These processes are called *transitions*. The energy of the photon emitted—which corresponds to the color of the light if it is in the visible range—is equal to the difference in energy between the two levels. Experiments have shown that a single energetic level is split in two if an external magnetic field is applied to the atom. As a consequence of such splitting, we can observe two transitions instead of a single one. *We observe two similar but distinct colors in the emitted light instead of one*. This phenomenon was explained by assuming that every electron carries a spin; therefore, the two levels would correspond to different orientations of the electron's spin. Different orientations contribute differently to the total energy of the electron. Since we are dealing with elementary particles, we have to use *quantum mechanics* to study the system, and, according to the laws of quantum mechanics, an electron spin can only have two orientations, and, thus, two levels. The electron *has* spin, but it does not spin.

See: Electricity, Light, Particle

THINK ABOUT IT: *Spectral Lines*

If it were possible to describe the electrons orbiting the nucleus of an atom in the same way we describe the motion of planets around the Sun, there would be a wide range of possible values for their energy according to their respective distances from the center. The electron's energy could have any value in that range. But this is not how the subatomic world works. According to the rules of *quantum mechanics*, the energy of the electron can have only certain fixed values. Further, every kind of atom has a different set of energy levels. This means that we can recognize an atom by studying its *spectral lines*—that is, the frequency of the light it emits when its electrons jump from one level to a lower one, since such frequency depends on the difference in energy of the two levels. Not only can we identify atoms in a lab by the light they emit, but we can study the light from stars to learn about the composition of distant galaxies.

READ ABOUT IT: *The History of the Physical World, The Basics, & Particle Spin*

A Brief History of Time [1988] by Stephen Hawking — This popular science book has seen its share of success. It remained on the *London Times* bestseller list for 237 weeks, and more than 9 million copies have been sold worldwide. Expect the book to be updated to include Hawking's 2004 admission that he was "wrong" about black holes.

Six Easy Pieces: Essentials of Physics Explained by Its Most Brilliant Teacher [1963] by Richard Feynman — This is another great book by Nobel-laureate Feynman, a leading 20th-century physicist and prolific author and lecturer, who always makes physics fun and compelling.

The Quark and the Jaguar: Adventures in the Simple and the Complex [1994] by Murray Gell-Mann — Yet another great book from yet another Nobel Prize-winning physicist. This one explores science concepts ranging from natural selection to computers to particle spin.

TALK ABOUT IT: *Beyond Physics—Spin & Synesthesia*

Spin is described as a measure of the electron's "sensitivity to a magnetic field" ... How are great artists similarly "sensitive to" and even "aligned with" such unseeable forces? ... Synesthesia is a neurological condition that involves the involuntary alignment of one sense with the perception of another (e.g., seeing sound, painting musically, counting in colors, finding personality in architecture) ... Is it possible that van Gogh could "feel" paint, or Beethoven "see" sound? ...

Wheatfield with Crows [1890] by Vincent van Gogh
Oil on canvas
Van Gogh Museum, Amsterdam

Wheatfield with Crows is so confident and so hauntingly familiar that one is left to marvel at the artist's comprehension rather than question the uniqueness of his vision. Van Gogh, who painted for only 10 years [1880-1890] reveals to the viewer something known but perhaps forgotten—a living, breathing world, one with a volatile, ecstatic underside, a world with *spin*. The overwhelming physicality of this work demonstrates an intense communion with nature. Looking at the painting, it is easy to feel as the artist must have: we are situated *there*, in the field, with the lowering birds and the shifting wheat, the tilting earth and the sky of many blues. The bright green road laid out before us is one on which we stand. The separateness from nature that we typically experience is no more than a formality of the mind. The viewer is reminded that although we can make the intellectual acquaintance of nature's elements—shape, color, position—we can only truly *know* sensation. Van Gogh's faith in the external world might have been a response to his interior disintegration: this is the last painting he completed before committing suicide.

"Águas de Março" ("Waters of March") [1972]
by Antonio Carlos Jobim

How to describe the indescribable? How to explain the hidden dynamism of life—the *spin*? Such questions move scientists *and* artists. There is no lovelier description of the cryptic subtlety of modern beauty than this bossa nova classic. The lyrics tick delicately like a melodic clock: *a stick, a stone*. Even industry as described here is not oppressive, but touching, and almost sad: *a truckload of bricks in the soft morning light*. "Águas de Março" reminds us that we are human, that we have a job to do, one for which we are uniquely outfitted—to witness, interpret, and render the beauty in life.

TORQUE 33

*Actively we have woven ourselves
with the very warp and woof of this nation,
—we fought their battles, shared their sorrow, mingled our
blood with theirs, and generation after generation have
pleaded with a headstrong, careless people to despise not
Justice, Mercy, and Truth, lest the nation be smitten with a
curse. Our song, our toil, our cheer, and warning have been
given to this nation in blood-brotherhood. Are not these gifts
worth the giving? Is not this work and striving? Would
America have been America without her Negro people?*

THE SOULS OF BLACK FOLK
W. E. B. DU BOIS

*The steer was down now, his neck stretched out,
his head twisted, he lay the way he had fallen.
Suddenly the bull left off and made for the other
steer which had been standing at the far end, his head
swinging, watching it all. The steer ran awkwardly
and the bull caught him, hooked him lightly
in the flank, and then turned away and looked up at
the crowd on the walls, his crest of muscle rising.
The steer came up to him and made as though
to nose at him and the bull hooked her perfunctorily.*

THE SUN ALSO RISES
ERNEST HEMINGWAY

From: **TORQUE** *n* (CIRCA 1884)
1 : A FORCE THAT PRODUCES OR TENDS TO PRODUCE ROTATION OR TORSION; *ALSO* : A MEASURE OF THE EFFECTIVENESS OF SUCH A FORCE THAT CONSISTS OF THE PRODUCT OF THE FORCE AND THE PERPENDICULAR DISTANCE FROM THE LINE OF ACTION OF THE FORCE TO THE AXIS OF ROTATION 2 : A TURNING OR TWISTING FORCE

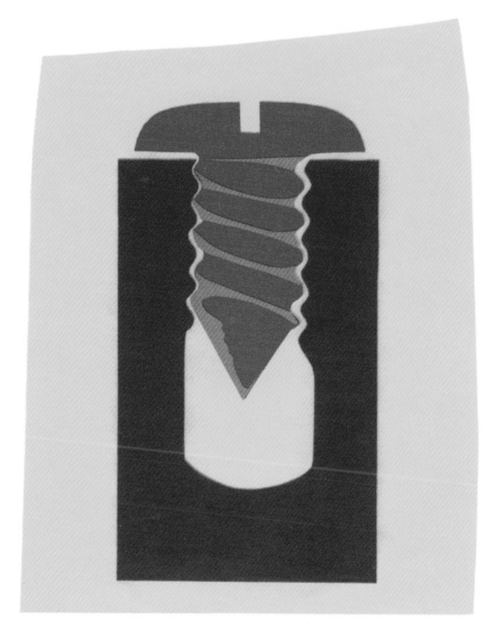

TORQUE

TORQUE IS THE PHYSICAL QUANTITY that describes the ways in which forces can make objects rotate about an axis. For example, we use a wrench to apply a torque to a bolt to turn it, and we apply a torque to a door when we open it. We know well that if we pull a door knob sideways, the door will not open. We would be applying a force to the door, but "zero" torque with respect to the door's axis of rotation (the side it is attached to the wall). In order to have a torque, the force we apply must be in the direction of rotation, or, more precisely, it must have a component along the direction of rotation.

Also, the farther away from the axis the force is applied, the greater the torque. The amount of torque needed to open a door can be given by a large force close to the axis or by a smaller force away from it. To open a revolving door with the least effort, we know to push it in a place farthest from the axis of rotation. In other words, though the *torque required to open the door* remains the same, the force required depends on the *distance from the axis of the point where it is applied*. A seesaw can be balanced by placing a heavier person closer to the center and a lighter one farther from center. The forces applied on the two sides are different because they are the respective weights of the bodies, but the contribution of each of the bodies to the total torque applied on the seesaw is *equal and opposite*, and so they cancel each other out. The larger weight is compensated by its smaller distance from the axis of rotation. The magnitude of the torque about a given axis depends on the magnitude of the force, its direction, and the location of the point at which the force is applied.

A torque is a vector. The direction of the vector representing the torque is the same as the direction of the axis of rotation. In the example of the opening door, the torque we apply is given by a vector aligned with the vertical axis, with length proportional to the magnitude of the torque, pointing up or down according to the direction of rotation. In the case of a simple screw, the direction of the torque we apply to it is, by convention, the same as the direction the screw will move when we turn it, and the direction is the same as the direction of vectors describing its angular velocity and angular acceleration.

***See:* Angle, Angular Velocity, Force, Position, Vector**

$$\tau = F d$$

[Eq. 11] ***The magnitude of the torque (τ) is equal to the product of the magnitude of the force (F) times the distance (d) of the axis of rotation to the line along which the force acts.***

THINK ABOUT IT: *Torque and Angular Acceleration*

Whenever a torque is applied on a rotating system, it determines an *angular acceleration*. Just as acceleration is the rate of change of velocity, angular acceleration is the rate of change of angular velocity: a torque makes the system rotate faster. Torque and angular acceleration are related by an equation analogous to the second law of dynamics. According to the second law, the acceleration a body experiences is proportional to the force acting on it and inversely proportional to its mass. For a *circular motion*, the angular acceleration is proportional to the torque and inversely proportional to the *moment of inertia* of the rotating body, which measures how the mass of the object is distributed with respect to the *axis of rotation*. The larger the moment of inertia, the harder it is to give the body an angular acceleration.

READ ABOUT IT: *Engineering & Scientific Discovery*

Remaking the World: Adventures in Engineering [1997] by Henry Petroski — Science and technology meet great prose in this collection of the author's essays from *American Scientist*. Petroski's writing is delightful. He discusses famous feats of engineering—the Eiffel Tower, the Hoover Dam, the Channel Tunnel, and the Ferris Wheel.

The Hole in the Universe [2001] by K. C. Cole — In terms both profound and poetic, Cole explores the idea of nothingness—the vacuum of space, the number zero—and how scientists, mathematicians, theologians, and philosophers have tried to grasp the concept.

The Structure of Scientific Revolutions [1962] by Thomas Kuhn — This book was named one of "the hundred most influential books since the Second World War," by the *Times Literary Supplement*. Like Hayden White's history of history, *Tropics of Discourse*, Kuhn's book takes a historical look at the process of scientific discovery and at established paradigms in order to examine whether legitimate challenges to prevailing notions and codes are even possible.

TALK ABOUT IT: *Beyond Physics—Torque & Language*

How is meaning established through language? ... Think of torque as "turning" language to apply the force of persuasion ... Consider the impact of irony and dialect on meaning ... Consider the conflict between accuracy and nuance in translation ... How is being explicit similar to being vague? ... Is leaving things open to interpretation a sign of trust, or is ambiguity actually a sign of mistrust? ...

***Jane Avril* [1899] by Henri de Toulouse-Lautrec**
Lithograph, printed in color
Musée Toulouse-Lautrec, Albi, France

It is hard to imagine an artist with a greater command of *torque* than Lautrec. The expressive linework of the diminutive genius who chronicled the enormity of a decadent Parisian society is singular and fantastic. Whether capturing the flare of a skirt, the undulating brim of a hat, or the intimidating crispness of a male silhouette at the Moulin Rouge, Lautrec could render with a twist of the wrist a sordidness in beauty, a loneliness in society, and a hollowness in joy.

***The Souls of Black Folk* [1903] by W. E. B. Du Bois**

This important collection of essays by the first African American to earn a Ph.D. from Harvard set into poetic and powerful terms the uniqueness of the American black experience of identity. To labor for a nation that does not respectfully regard you as its own is an example of citizenship with a bitter twist—*torque*.

***The Sun Also Rises* [1926] by Ernest Hemingway**

Hemingway wrote with a lean style often referred to as *masculine*; and yet manhood, as he described it, was complex. Here, the idealized strength of bullfighting stands in stark contrast to the pathos of the war-damaged protagonist unable to consummate a relationship with the woman he loves. Like the bull who twists the horn to make the kill, Hemingway writes with *torque*, forcing readers with language infused with tension to reassess expectations of potency.

UNCERTAINTY 34

To be, or not to be—that is the question:
Whether 'tis nobler in the mind to suffer
The slings and arrows of outrageous fortune
Or to take arms against a sea of troubles
And by opposing end them. To die, to sleep
No more, and by a sleep to say we end
The heartache, and the thousand natural shocks
That flesh is heir to. 'Tis a consummation
Devoutly to be wished. To die, to sleep;
To sleep—perchance to dream—ay, there's the rub,
For in that sleep of death what dreams may come
When we have shuffled off this mortal coil
Must give us pause. There's the respect
That makes calamity of so long life.
For who would bear the whips and scorns of time,
Th' oppressor's wrong, the proud man's contumely,
The pangs of despised love, the law's delay,
The insolence of office, and the spurns
That patient merit of th' unworthy takes,
When he himself might his quietus make
With a bare bodkin? Who would fardels bear,
To grunt and sweat under a weary life,
But that the dread of something after death,
The undiscovered country, from whose bourn
No traveler returns, puzzles the will,
And makes us rather bear those ills we have
Than fly to others that we know not of?
Thus conscience does make cowards of us all,
And thus the native hue of resolution
Is sicklied o'er with the pale cast of thought,
And enterprises of great pitch and moment
With this regard their currents turn awry
And lose the name of action.

HAMLET, ACT III, SCENE 1
WILLIAM SHAKESPEARE

From: **UN-CER-TAIN-TY** *n* (1929)
THE UNCERTAINTY PRINCIPLE STATES NOT THAT THE OBSERVER ALWAYS INTERFERES WITH THE OBSERVED, BUT RATHER THAT AT A VERY FINE GRAIN SIZE, THE WAVE-PARTICLE DUALITY OF THE MEASURING TOOLS BECOMES RELEVANT.[18]

UNCERTAINTY

EVERY MEASUREMENT OF PHYSICAL QUANTITIES comes with a definite uncertainty—a number that estimates how far from the "true value" we might be. If we measure the length of a pen with an ordinary ruler, we cannot easily measure differences in length that are smaller than 1 millimeter. We can express the *uncertainty* on our measurement by saying that the pen is 130 millimeters long, plus or minus 1 millimeter. One millimeter is our uncertainty.

In comparing the pen to the ruler, we might think that nothing touches the pen, but, in fact, something does touch the pen—a ray of light is reflected by it. That is how we see the pen. We do not care how much the light affects the position of the pen itself. If the light does affect the position, it must obviously be less than our uncertainty on the measurement—again, 1 millimeter. But if we need to measure the size of one of the pen's atoms, or the position of a single electron in one of its atoms, light would affect the state of the system in a relevant way. However, it is the only way: we can only study particles by bouncing something off them. The problem is that if we want to determine the position of an electron with a small uncertainty, we need to use light that has a high frequency, that is, a short wavelength, since the shorter the wavelength, the better the resolution.

At this point, quantum mechanics enters the picture. Light comes in photons, packets (*quanta*) of energy. The higher the frequency of the light beam, the larger the energy each photon carries. When the high-energy photon interacts with the electron, it changes the momentum of the electron in an unpredictable way. If we use less energetic photons, we lose information on position. This relation between the *uncertainties on position and the momentum of a particle* is not a problem of measurement. It explains the nature of the subatomic world, and so it became a fundamental principle of quantum mechanics, the theory describing the interaction of elementary particles. The uncertainty principle states that if we want to determine the position of a particle with a small uncertainty, then we must expect a large uncertainty on its momentum. In quantum mechanics, concepts such as the trajectory of a particle have no meaning.

***See:* Light, Measurement, Momentum, Particle, Position**

THINK ABOUT IT: ***Classical and Quantum Mechanics***

In *classical mechanics*, the theory based on Newton's laws of motion, the typical problem consists of finding the trajectory a body follows after determining its initial position and velocity, as well as the forces acting on it. In principle, the trajectory can be predicted with infinite accuracy. In *quantum mechanics*, things are completely different: uncertainty is part of the theory. We cannot predict a trajectory for a particle because that would imply that we could know—exactly—the particle's position. But that's not true, even in principle. The theory can predict probabilities. For example, it can predict the probability of finding the particle in a certain region of space. In determining the probability for different outcomes of lab measurements, quantum mechanics proves to be extremely successful.

READ ABOUT IT: ***Uncertainty, Math, & The Mind***

From Certainty to Uncertainty [2002] by F. David Peat — Peat frames his discussion of physics in relation to the transformation from the "certainty" that categorized approaches to science and life in the 19th century to the "uncertainty" that marked the 20th century. The book is separated into brief and clearly delineated sections concerning such things as Impressionism, antibiotics, environmentalism, risk analysis, etc. Above all, this book celebrates the joys of curiosity and discovery.

Metamagical Themas: Questing for the Essence of Mind and Pattern [1985] by Douglas Hofstadter — This giant and informative collection of quirky and approachable essays contains an entire chapter devoted to Heisenberg and the uncertainty principle.

TALK ABOUT IT: ***Beyond Physics—Uncertainty & Value***

Uncertainty in physics and uncertainty in modern life ... As our "tools" for measuring the world become more refined, how does our uncertainty become "larger"? ... Can we measure without affecting the subject of our measurement? ... By what means can we assess "true value"? ... What does it mean to "trust the moment," "leave things to chance"? ... How are the spontaneously-created forms of nature (a rose) more universally understood and appreciated than manufactured ones (a language), which are inseparable from the culture in which they were made? ... In what ways does nature place a premium on spontaneity, impulsiveness, instinct—uncertainty? ... Consider falling in love versus an arranged marriage ... Is the information derived from "measured" circumstances more certain than the information derived from "unmeasured," spontaneous ones? ...

***The Green Trees* or *Beech Trees in Kerduel* [1893]**
by Maurice Denis
Oil on canvas
Musée d'Orsay, Paris

Denis, who studied at the École des Beaux-Arts, was a founding member of the Nabis, a post-Impressionist artistic fraternity based in Pont-Aven, Brittany. The members of the Nabis were influenced by Gauguin's (See: *Friction*) pictorial symbolism and expressionistic use of color. There is a geometry to *The Green Trees* that is strong and secure, but also sinuous and eerie. Beyond the rhythmic pattern of the trees lies a subjective, experiential, *uncertain* world—one that defies definition, one that changes the moment it is rendered.

***Hamlet* [1601] by William Shakespeare**

To be or not to be? In this legendary examination of *uncertainty*, Hamlet questions the value of *action* (i.e., avenging the murder of his father), and Shakespeare probes the process of *becoming* (i.e., will Hamlet remain the son, or will he become a man—the next father and king—through that action?). Alas, Hamlet succeeds best only at questioning. When we meet him, he is a brooding, self-centered prince, cognizant of—but ambivalent about—duty and revenge, and by the play's end, he is unchanged. Ironically, he dies at the hand of Laertes, who has no problem avenging the death of *his* own father, Polonius. At last, in fulfillment of his soliloquy, (*To die, to sleep/No more, and by a sleep to say we end/The heartache, and the thousand natural shocks/That flesh is heir to*), Hamlet is extinguished.

VECTOR 35

Like an expert singer skilled at lyre and song—
who strains a string to a new peg with ease,
making the pliant sheep-gut fast at either end—
so with his virtuoso ease Odysseus strung
his mighty bow. Quickly his right hand
plucked the string to test its pitch
and under his touch it sang out clear
and sharp as a swallow's cry.
Horror swept through the suitors, faces blanching
white, and Zeus cracked the sky with a bolt,
his blazing sign, and the great man who
had borne so much rejoiced at last
that the son of cunning Cronus
flung that omen down for him.
He snatched a winged arrow lying bare
on the board—the rest still bristled deep
inside the quiver, soon to be tasted by all
the feasters there. Setting shaft on the handgrip,
drawing the notch and bowstring back, back ...
right from his stool, just as he sat but
aiming straight and true, he let fly—
and never missing an ax from the first ax-handle
clean on through to the last and out the shaft
with its weighted brazen head shot free!

THE ODYSSEY
[BOOK 21: ODYSSEUS STRINGS HIS BOW]
HOMER
TRANSLATED BY ROBERT FAGLES

From: **VEC-TOR** *n* (1846)
1 A : A QUANTITY THAT HAS MAGNITUDE AND DIRECTION AND THAT IS COMMONLY REPRESENTED BY A DIRECTED LINE SEGMENT WHOSE LENGTH REPRESENTS THE MAGNITUDE AND WHOSE ORIENTATION IN SPACE REPRESENTS THE DIRECTION B : A COURSE OR COMPASS DIRECTION ESPECIALLY OF AN AIRPLANE 2 A : AN ORGANISM (AS AN INSECT) THAT TRANSMITS A PATHOGEN B : POLLINATOR

VECTOR

A *VECTOR* IS A MATHEMATICAL tool used to describe physical quantities that have both magnitude and direction. Because we live in a three-dimensional world in which objects can move in many directions, vectors are used to describe quantities such as position, velocity, acceleration, and force. If we say that an apple is falling at 20 meters per second, we are giving the speed of the apple and its direction of motion. A vector representing the apple's velocity can be drawn as an arrow pointing downward. The length of the arrow would be proportional to the speed. The vector representing the velocity of the apple at some later time when the speed increases would be drawn as an arrow that is proportionally longer. A vector has three components: three numbers are needed to describe all of its possible orientations, as well as its magnitude. For example, a vector can describe the position of Mexico City with respect to New York City. Its components would be how "much" south, how "much" west, and how "much" high (Mexico City is 2000 meters above sea level), we need to travel to get from one place to the other.

Vectors also describe forces. In order to know how a force is affecting a physical system, we need to know its intensity and its direction. If more than one force is acting on a body at the same time, we need to combine their effects properly, and we need to compute the resulting total force acting on the body—we need to *sum* the vectors. In the case of a pendulum, there are two forces: the gravitational force, which always points downward, and the tension of the string to which it is attached, which always points to the place the pendulum hangs. When a pendulum is at its lowest position, the two forces are *equal in intensity* but *opposite in direction*. The sum of the two vectors describing them is zero; therefore, the total force on the pendulum is zero. The lowest position is, for the pendulum, its position of stable equilibrium. As the pendulum moves to the left, the tension of the string is no longer vertical, and it will be described by a vector with a certain horizontal component. The gravitational force that always has only a vertical component cannot balance it anymore. The total force will be directed to the right and will tend to bring the pendulum back to its equilibrium position.

See: ***Acceleration, Angle, Angular Velocity, Force, Position, Velocity***

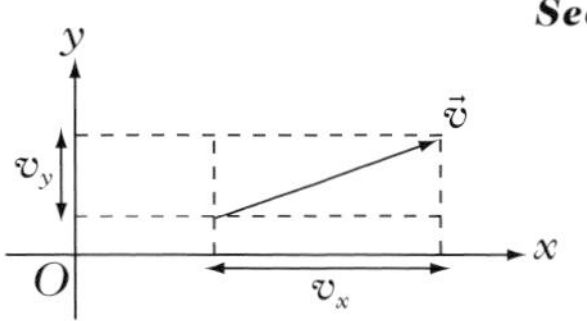

[Ill. 5] The description of the vector ($\vec{v}$) on a plane in Cartesian coordinates. The components of the vector in the direction of the x- and y-axes are v_x and v_y.

THINK ABOUT IT: *Vectors and Scalars*

Vectors describe many different quantities in physics. Both electric and magnetic fields are *vector fields* because they have magnitude and direction at every point in space. A compass needle orients itself along the lines of the Earth's magnetic field: the direction in which the needle points corresponds to the direction of the magnetic field. On the other hand, quantities that can be represented by a single number are called *scalar quantities*. The mass and the energy of a body are scalars, not vectors.

READ ABOUT IT: *Time Travel, Useful Objects, & The Workings of the Mind*

The End of Time: The Next Revolution in Physics [1999] by Julian Barbour — Barbour questions all the things we *think* we know about the phenomenon that is time. He explores themes of time travel, parallel universes, illusion, and mortality. In his discussion of wave patterns, he refers to Shakespeare's *A Midsummer Night's Dream*.

The Evolution of Useful Things [1992] by Henry Petroski — Petroski investigates in detail the evolution of household artifacts. Chapters include "Patterns of Proliferation" and "When Good is Better than Best." "Inventors as Critics" marks a great starting point for a discussion about turning life's annoyances into worthy inventions.

How the Mind Works [1997] by Steven Pinker — Whether one reads this book in its entirety or selects chapters at random, one will be indelibly enriched. Linguist and cognitive scientist Pinker discusses the ways the brain processes and interprets information. One section covers the evolutionary basis for emotions and the difference between "real" and "sham" emotions, while another, "Revenge of the Nerds," discusses cognitive ability in humans versus other species.

TALK ABOUT IT: *Beyond Physics—Vector & Destiny*

Destiny is thought to be a predetermined course of events to which a person or thing is held ... How is individual destiny like a vector? ... Is it possible that for every person there is one right path? ... How is each of us supposed to find his or her "way"? ... To what extent is one's destiny based on unavoidable and externally occurring events, and to what extent is it based on unique, internal responses to such externally occurring events? ... What does it mean when people say, "it was destined to happen this way"? ... Why is it tempting to attribute exceptional events to destiny than to pure chance? ...

Marat assassiné (The Death of Marat) **[1793]**
by Jacques-Louis David
Oil on canvas
Musées Royaux des Beaux-Arts de Belgique, Brussels

"Live by the sword, die by the sword." Marat, a leader of the French Revolution and a man of ideas and ideals, was killed in his bathtub by an enemy of equally potent sentiments. In the foreground, beneath his elbow, rests the knife that Royalist sympathizer Charlotte Corday plunged into his chest. David's neoclassical rendering of the death of a martyr is ironically tranquil. Marat, who had been writing on a board in his bath, is inundated by pure white light and framed by luxurious green fabric, his corpse not yet caught in rigor mortis. Here, *vector* can be seen as descriptive not only of the blade used to kill, but of the passion of ideals and the consequence of the desire for change—revolution itself.

The Odyssey **[8th-7th century B.C.E.] by Homer**
Translated by Robert Fagles

The concept of *vector* recalls one of the most thrilling moments in narrative fiction: Odysseus, dressed as a vagrant, triumphantly shoots an arrow through the holes of twelve lined-up ax heads to win back the hand of his own wife. After ignoring the menacing jeers of the suitors who abused his home and pillaged his larder during his 20-year absence, Odysseus, along with his son, Telemachus, kills them all before revealing himself to the faithful Penelope.

VELOCITY 36

I wasn't the kind of guy
who ran from just anything; I was going to know
who or what I was running from. So I stood there
for a while. Turk started running. Then I heard
a shot, one shot. Blam! Then I saw some fire
from a gun, and I started running. When I got about
midway on the stairs leading up from the backyard,
it seemed to just dawn on me. I said, "Oh, shit,"
somebody's shootin' at us." I kept on running.
Then, after I had gotten out of the backyard,
I don't know what happened but suddenly I knew
I was shot. I didn't feel any blood right away;
I didn't feel any pain; I didn't feel anything.
All I felt was that I was slowing down. It was like
something had a hold of me, and I knew it was a bullet.

MANCHILD IN THE PROMISED LAND
CLAUDE BROWN

From: **VE-LOC-I-TY** *n* (CIRCA 1550)
1 A : QUICKNESS OF MOTION: SPEED <THE ~ OF SOUND> B : RAPIDITY OF MOVEMENT C : SPEED IMPARTED TO SOMETHING 2 : THE RATE OF CHANGE OF POSITION ALONG A STRAIGHT LINE WITH RESPECT TO TIME : THE DERIVATIVE OF POSITION WITH RESPECT TO TIME 3 A : RATE OF OCCURRENCE OR ACTION : RAPIDITY

VELOCITY

VELOCITY IS THE MEASURE OF the rate of change in an object's position. The velocity of an object is given not only by its speed—how fast it moves with respect to something else—but by its direction of motion. For this reason, velocity is a vector quantity.

If we are interested in knowing the speed of a car traveling on a highway, we can determine it by measuring the distance covered along the highway in a certain interval of time. If the car covers 10 miles in 10 minutes, we can say that its average speed during that time is one mile per minute, or 60 miles per hour. If the car travels 10 miles in 5 minutes, or 20 miles in 10 minutes, the average speed is doubled. We call this quantity *average speed* because during the course of a 100-mile road trip, a given car might slow down or stop several times due to such factors as weather or traffic conditions. But if we know that the 100-mile trip took two hours overall, then we can calculate, in retrospect, that the average speed was 50 miles per hour. What was changing during the trip was what is called the *instantaneous speed*. The instantaneous speed is calculated in the same way as the average speed, but it considers very small, infinitesimal changes of position over infinitesimal intervals of time.

In order to describe the motion of objects in our three-dimensional world, we need to be able to express their speed as well as their direction of motion. The velocity is a vector always pointing along the direction of motion. The direction of motion of an airplane coming down for landing is as important as its speed. The pilot knows that the velocity of the plane has to make a precise angle with the horizontal ground. Mathematically, the velocity of the plane can be described in terms of two components: a horizontal component and a vertical component. The vertical component of the velocity would then be a measure of the rate of change of the altitude of the plane. For instance, the plane's velocity possibly could have a horizontal component of 200 miles per hour and a vertical one of 10 miles per hour. Neither of these numbers is the actual speed of the plane, but if they were to represent the two sides of a rectangle, the actual speed would be the rectangle's diagonal.

***See:* Acceleration, Equation, Measurement, Position, Vector**

$$\vec{v} = \frac{\Delta \vec{x}}{\Delta t}$$

[Eq. 12] ***The average velocity ($\vec{v}$) of a body in a time Δt is given by the ratio of the change in position ($\Delta \vec{x} = \vec{x}_{final} - \vec{x}_{initial}$) with the time elapsed ($\Delta t = t_{final} - t_{initial}$). The symbol Δ in front of a physical quantity often indicates how much such quantity changes, a difference between final and initial values.***

THINK ABOUT IT: *Velocity & Trajectory*

The velocity of a body is a *vector* that points in the direction of the motion of the body. If we move along a circle at constant speed, our velocity never changes in *magnitude*, but it constantly changes in *direction*. In order to draw the vector representing our velocity in a given place, we would draw an arrow starting from that point and tangent (adjacent) to the circle. If we swing a rock with a string, and the string breaks at a certain time, the rock will move in the direction of the velocity it had at the time the string broke, a direction tangent to the original circle. Even in the case of more complicated motions, the velocity of a body is a vector that is always tangent to the trajectory of the body.

READ ABOUT IT: *A Two-Dimensional World, Alternative Energy Sources, Vacuums, & Skateboarding*

Flatland: A Romance of Many Dimensions [1884] by Edwin Abbott — The characters in this parody of Victorian society are geometric shapes who cannot comprehend anything other than the two-dimensional world they occupy. Eventually, the narrator, "A. Square," is thrust into the land of three-dimensions. Notably, Abbott probes the era's irrational social attitudes, particularly those concerning women.

Tomorrow's Energy: Hydrogen, Fuel Cells, and the Prospects for a Cleaner Planet [2002] by Peter Hoffman — Most of the fuels humans use to power industrial and household machines result in seriously harmful effects to the environment. Hoffman suggests that the answer to this dilemma might be found in developing the use of hydrogen, the most abundant element in the universe.

Dogtown and Z-Boys [2001] directed by Stacy Peralta — This high-velocity skateboarding film features classic footage from the 1970s. Watch it to enjoy some "applied dynamics." Narrated by Sean Penn.

TALK ABOUT IT: *Beyond Physics—Velocity & Social Determinism*

How can the concept of velocity be applied to the "rate and direction" of life events? ... Depending upon time and place, events can follow one another too rapidly for human assessment ... Even if there is the will or desire to change course, often there is no opportunity to do so ... Social determinism is a hypothesis suggesting that social constructs and social exchanges alone determine human behavior ... What are the limitations of such a philosophy? ...

***No. 7* [1970] by Robert Motherwell**
from *The Basque Suite*
Screenprint on paper
Tate Gallery, London

This pioneer of abstract expressionism was a teacher, lecturer, editor, and student of philosophy and art history at Stanford and Harvard, as well as the husband of artist Helen Frankenthaler (See: *Field*). Though *No. 7* is an abstract shape, it is suggestive and gently moving. Because the image is not literal, the viewer can envision *anything*: for instance, the bold black and blue horizontal marks leading off to the lower right can be thought of as lines indicating movement, the upright geometric A-shape as a gliding sailboat, and the irregular edges as indicative of undulating water. Seen this way, *No. 7* is like a single frame of film depicting a boat's motion, its *velocity* trapped in time.

***Manchild in the Promised Land* [1965] by Claude Brown**

In his chronicle of the perils and the blessings of being a young black man in Harlem in the 1950s, Brown gives us a moving autobiography and an intense historical record. Familiar language rips across the pages like a high-speed train, but the overall tone is poignant, even nostalgic. In this excerpt, Sonny gets shot but he keeps on running. The reader bears witness to the *velocity* of living and to what would have been, for the main character, the almost certain danger of slowing down and thinking things through.

WAVE

37

Very Simple Slogans
1. Film-drama is the opium of the people.
2. Down with the immortal
kings and queens of the screen! Long live the ordinary
mortal, filmed in life at his daily tasks!
3. Down with the bourgeois fairy-tale script!
Long live life as it is!
4. Film-drama and religion are deadly weapons
in the hands of the capitalists. By showing our
revolutionary way of life, we will wrest
that weapon from the enemy's hands.
5. The contemporary artistic drama
is a vestige of the old world. It is an attempt to pour
our revolutionary reality into bourgeois molds.
6. Down with the staging of everyday life!
Film us as we are.
7. The scenario is a fairy-tale
invented for us by a writer. We live our own
lives, and we do not submit to anyone's fictions.
8. Each of us does his task in life and does not
prevent anyone else from working. The film workers' task is
to film us so not to interfere with our work.
9. Long live the kino-eye of the proletarian revolution!

KINO-EYE: THE WRITINGS OF DZIGA VERTOV
DZIGA VERTOV
TRANSLATED BY KEVIN O'BRIEN
EDITED BY ANNETTE MICHELSON

From: **WAVE** *n* (1526)
1 A : A MOVING RIDGE OR SWELL ON THE SURFACE OF A LIQUID 2 A : A SHAPE OR OUTLINE HAVING SUCCESSIVE CURVES C : AN UNDULATING LINE OR STREAK OR A PATTERN FORMED BY SUCH LINES 3 : SOMETHING THAT SWELLS AND DIES AWAY : AS A : A SURGE OF SENSATION OR EMOTION B : A MOVEMENT SWEEPING LARGE NUMBERS IN A COMMON DIRECTION <~*S* OF PROTEST> 7 A : A DISTURBANCE OR VARIATION THAT TRANSFERS ENERGY PROGRESSIVELY FROM POINT TO POINT IN A MEDIUM AND THAT MAY TAKE THE FORM OF AN ELASTIC DEFORMATION OR OF A VARIATION OF PRESSURE, ELECTRIC OR MAGNETIC INTENSITY, ELECTRIC POTENTIAL, OR TEMPERATURE B : ONE COMPLETE CYCLE OF SUCH A DISTURBANCE

WAVE

IMAGINE TOSSING A PEBBLE INTO a still lake. It strikes the surface, and we observe circles, or rings, radiating outward. The molecules on the water's surface are simply moving up and down, and yet, there is something traveling with a given speed, some pattern in the water. That traveling pattern is called a *wave*, and the water in which it travels is the medium. If it's hard to imagine all the tiny molecules that comprise a big lake moving up and down, think of sports fans in a stadium "doing the wave." Nobody leaves their position, yet the wave travels across the stadium. In this case, the medium is the people. Both the waves—in the lake and in the stadium—are *transverse waves* because the up-and-down movement of the molecules/people is *perpendicular* to the direction of propagation of the wave.

Sound waves are an example of a *longitudinal wave*. As music plays, the speakers vibrate, shaking the air back and forth, making a pattern of waves that reaches our ears as sound. The air molecules are not really moving through space any more than the water molecules were. The important difference is that the air shakes in the direction of the propagation of the sound wave, not perpendicular to it. Light is also a wave but different from the examples we have seen before because it does not need a medium to propagate. Light is an electromagnetic wave; it consists of oscillating electric and magnetic fields.

Important properties of a wave are its *frequency*, *wavelength*, *speed*, and *amplitude*. The frequency of a wave is the number of times the medium oscillates per unit of time. If we were to float a rubber duck on a lake, the frequency of the waves is indicated by how often the duck goes up and down. In music, the frequency of a sound wave is called its pitch. The wavelength is the distance between the closest peaks. In the ocean, it is the distance between the tops of two successive waves. The speed of propagation of a wave is equal to the product of its frequency and its wavelength. Amplitude refers to the height of the peak. Thinking about the rubber duck on water waves, the amplitude measures the height the duck reaches when it is at the top of the wave. With sound, the amplitude is the volume, or the loudness, of the sound.

See:* *Reflection, Refraction

$$v = \lambda f$$

[Eq. 13] ***The speed of propagation of a wave (v) is equal to the product of its wavelength (λ) and its frequency (f).***

THINK ABOUT IT: *Making Waves on Ropes*

Imagine a wave produced by snapping the end of a rope—we can see a crest move up and down along the length of rope. Although the rope itself is not moving forward, the wave we created is, and the energy we used to snap the end of the rope also moves from one side of the rope to the other. The higher we pull the end, the greater the *amplitude* of the wave, the larger the energy. Waves transport energy without transporting matter. Notice that no matter how fast we move the end of the rope up and down, the speed at which the wave *propagates* does not change. The speed depends only on the particular rope we use. In general, the speed of propagation of a wave depends only on the *medium* in which it propagates (e.g., water, air).

READ ABOUT IT: *Physics, Philosophy, Quantum Theory, & The Corporate Model*

Physics and Philosophy [1943] by Sir James Jeans — This excellent book by Jeans connects the history and the theory of two separate disciplines. In chapters such as "Religion and Science," "Mentalism or Materialism," the author examines *a priori* knowledge, modes of thought, idiom, causality, determinism, free will, fate, etc.

Thirty Years that Shook Physics: The Story of Quantum Theory [1966] by George Gamow — Gamow humorously discusses the experiences and the discoveries of friends and colleagues (e.g., Niels Bohr, Max Planck, Ernest Rutherford, Wolfgang Pauli, et al.). The author challenges the traditional learning model for science by teaching with humor, caricature, anecdotes, and dramatic plays.

The Corporation [2004] by Mark Achbar, Jennifer Abbott, and Joel Bakan — This award-winning documentary offers an unforgettable view of contemporary corporate philosophy. Is the corporate model at the crest of its power or will it fade in significance? Is it a sustainable method of business? Is it providing for future generations?

TALK ABOUT IT: *Beyond Physics—Waves & Innovation*

How do ideas advance in waves? ... How does human innovation "move in a common direction"? ... What is the importance of public readiness to the acceptance of new products? ... What is the function of technological inventions—vanity, well-being, commerce? ... Do they always improve the conditions of living or do they ever complicate them unnecessarily? ... What do waves teach us about time? ... How are trends like surging tides—swelling then dying away? ...

***The Migration of the Negro, Panel 3* [1940-1941]**
by Jacob Lawrence, *From Every Southern Town Migrants Left by the Hundreds to Travel North*
Tempera on Masonite
The Phillips Collection, Washington, D.C.

For African Americans of the early 1900s, the North held the promise of opportunities for work and self-actualization. Lawrence depicts those who moved on from the legacy of slavery to the hope of better living during the Great Migration. Spurred by enthusiastic letters from friends and relatives and by reports in the black-owned *Chicago Defender* newspaper, a *wave* of 1.5 million people migrated (like the birds in the painting) from the segregated South. Note that the barren landscape the migrants abandon is enriched—fleetingly, ironically—only by the departing figures. Lawrence is one of the most famous members of the Harlem Renaissance [1920s], an artistic and cultural movement that was stifled by the economic crisis of the Depression.

***Kino-Eye: The Writings of Dziga Vertov* [1926]**
by Dziga Vertov
Translated by Kevin O'Brien, Edited by Annette Michelson

The Communist ideology as professed here might seem excessive to contemporary readers, but history is rich with examples of regimes in which the political extremism of those in authority led to the rise of political extremism in the citizenry. In fact, counter-extremism is in many cases a matter of life and death. Whether or not individuals actually ascribe to intolerance in their hearts, or just in language, such movements are best seen in historical context, perhaps as *waves*—rising, cresting, and eventually receding. Vertov produced 23 issues of his *Kinopravda* film-newspaper [1922-1925].

WAVE/PARTICLE DUALITY 38

Light behaves like waves or like particles depending upon which experiment we perform. The "we" that does the experimenting is the common link that connects light as particles and light as waves. The wave-like behavior that we observe in the double-slit experiment is not a property of light, it is a property of our interaction with light. Similarly, the particle-like characteristics that we observe in the photoelectric effect are ... a property of our interaction with light.

Wave-like behavior and particle-like behavior are properties of interactions. Since particle-like behavior and wave-like behavior are the only properties that we ascribe to light, and since these properties now are recognized to belong ... not to light itself, but to our interaction with light, then it appears that light has no properties independent of us! To say that something has no properties is the same as saying that it does not exist. The next step in this logic is inescapable.

Without us, light does not exist. Transferring the properties that we usually ascribe to light to our interaction with light deprives light of an independent existence. Without us, or by implication, anything else to interact with, light does not exist. This remarkable conclusion is only half the story. The other half is that, in a similar manner, without light, or, by implication, anything else to interact with, we do not exist!

THE DANCING WU LI MASTERS
GARY ZUKAV

From: **WAVE / PAR-TI-CLE DU-AL-I-TY** (20C)
THE CONCEPT IN QUANTUM MECHANICS THAT THERE IS NO DISTINCTION BETWEEN WAVES AND PARTICLES; PARTICLES MAY SOMETIMES BEHAVE LIKE WAVES, AND WAVES MAY SOMETIMES BEHAVE LIKE PARTICLES.[19]

WAVE/PARTICLE DUALITY

In classical physics, the distinction between particle and wave is very clear, and it corresponds to what is suggested by common sense. The word *particle* implies something localized that moves along a clear path. If an electron moves quickly through some material, it can leave a trail that can be observed in certain experiments. The trail provides evidence that a little bunch of "something" passed through some definite spots. That is particle behavior.

The thing that marks the difference between a wave and a particle is a phenomenon called *interference*. Only waves can produce interference. The simplest example of interference is observed when two pebbles are thrown into a pond at the same time. A single pebble would produce waves propagating out toward the edge in rings. But the rings produced by two pebbles would meet, overlap, and form something similar to a geometrically regular web of ups and downs in which the original rings could hardly be recognized. Two wave peaks can add up at some points, while a peak and a valley can add up—to zero—at others. We say that the two waves give rise to spots of *constructive* and *destructive* interference. Two beams of light of the same frequency can produce a similar effect. In classical physics, light is described as an *electromagnetic wave*.

It is very difficult to imagine two beams of particles producing destructive interference—this would mean that the two beams meet at a point at which there are no particles. Nevertheless, interference between two beams of electrons has been observed in particular experiments. Apparently, electrons do behave like waves. Also, we now know that light can interact like a particle, not just like a wave. If we shine a very dim light on a very sensitive piece of film and watch the pattern as it develops, we can see lots of tiny dots arriving one at a time, because the light hits the film in little bunches, rather than as one even exposure. The old theory, with its clear distinction between particles and waves, could not describe these observations. Quantum mechanics explained these strange behaviors, but in order to do this, it had to abandon the categories of "particle" and "wave" of classical physics.

***See:* Energy, Light, Particle, Wave**

THINK ABOUT IT: ***Quantum, Again***

The behavior of particles at a microscopic level is different from the behavior of macroscopic objects with which we have direct experience. The theory that describes such particle behavior, *quantum mechanics*, is not intuitive; however, it makes mathematical sense. *Erwin Schrödinger's* [1887-1961] equation is the fundamental equation of quantum mechanics. It plays the role that Newton's second law of dynamics does in classical physics. The description that the Schrödinger equation gives of microscopic phenomena is correct in the sense that experiments show what the theory predicts. In physics, that's all that matters.

READ ABOUT IT: ***Signs and Language, The Civil War, & Ancient Eastern Thought***

Mythologies [1957] by Roland Barthes — The nature of words and their meaning is examined by Barthes in this charming collection of essays ("The Jet-Man," "The Brain of Einstein," "Toys," and "Plastic"). In "Steak and Chips," steak is *more than* steak—it is a *sign* of virility and dominance. Like fellow literary and cultural theorists, Barthes evaluated systems of ideas to tease apart the ways in which meaning is made, conveyed, and shared.

"An Occurrence at Owl Creek Bridge" [1909] by Ambrose Bierce — *Is he dead or is he alive*? This creepy story about the Confederate family man really leaves you hanging—quite literally! Like many of Bierce's stories, "Owl Creek" is dark and menacing, but also provocative. Bierce fought in some of the Civil War's heaviest battles, and his work reflects the fact that he witnessed the dual nature of mankind.

Tao Te Ching [circa 500 B.C.E.] by Lao Tzu, translated by John Wu — This classic text has had great influence on Eastern *and* Western thought. It offers a view on the duality of something other than "waves and particles" (e.g., of humankind, the universe, and life).

TALK ABOUT IT: ***Beyond Physics—Duality & Unity***

Consider duality ("either/or-ness") versus unity ("both-ness") ... When should completion be the goal, and when is perfect opposition enough? ... Night/day, clarity/ambiguity, matter/spirit, celestial/terrestrial, positive/negative, emptiness/fullness, yin/yang ... How are such correlatives indivisible? ... Can right exist without wrong, peace without war, hope without despair? ... Are there any concepts important to human life that do not have opposites (e.g., grief)? ...

Une Étoile caresse le sein d'une négresse
***(peinture-poème)* [1938] by Joan Miró**
Oil on canvas
Tate Gallery, London

Is it two triangles or a woman? A poem or a painting? A ladder or an easel? The answer to such questions—*it's both, depending on how you look at it*—is similar to the one used by physicists to describe the *duality* of light (i.e., Is it a *wave* or a *particle*?). Miró, one of the most lyrical of the modern surrealist artists, produced work that was as serious and symbolic as it was inventive and spatially free. Artists like Miró, Magritte (See: *Relativity*), and Salvador Dalí celebrated the primacy and the magical quality of dreams, considering them to be the best avenues to the riches of the subconscious.[41]

***The Dancing Wu Li Masters* [1979] by Gary Zukav**

Though it would have been difficult to find a literary passage that elucidates *wave/particle duality*, it probably would not have been impossible. However, Zukav's description of wave-like behavior and particle-like behavior as properties of *human interaction with light* is so lucid and poetic that it warranted inclusion. The duality principle as so described by Zukav (*Without us light does not exist*) is reminiscent of the question: if a tree falls in a forest, and there is no person there to hear it, does it make a sound? The next logical question might be: what about the animals who hear it? Or in the case of light, the animals who *see* it? We are reminded that though the *principles* of physics are integral to daily living, *descriptions* of physics often make the subject seem daunting. *Wu Li* succeeds at opening minds to the wonder and joy of science.

All old work nearly has been
hard work. It may be the hard work of children,
of barbarians, of rustics; but it is always their utmost.
Ours has as constantly the look of money's worth, of a
stopping short wherever and whenever we can, of a lazy
compliance with low conditions; never of a fair putting
forth of our strength. Let us have done with this kind
of work at once: cast off every temptation to it: do not
let us degrade ourselves voluntarily, and then mutter and
mourn over our shortcomings; let us confess our poverty
or our parsimony, but not belie our human intellect.

THE LAMP OF BEAUTY
JOHN RUSKIN

After the statue had been finished,
its great size provoked endless disputes over the best way to
transport it to the Piazza della Signoria. However, Giuliano
da Sangallo, with his brother Antonio, constructed a very
strong wooden framework and suspended the statue from
it with ropes so that when moved it would sway gently
without being broken; then they drew it along by means of
winches over planks laid on the ground, and put it in place.
In the rope which held the figure suspended he tied a slip-
knot which tightened as the weight increased: a beautiful
and ingenious arrangement.... And without any doubt this
figure has put in the shade every other statue, ancient or
modern, Greek or Roman. Neither the Marforio in Rome,
nor the Tiber and the Nile of the Belvedere, nor
the colossal statues of Monte Cavello can be compared
with Michelangelo's David.... The grace of this figure
and the serenity of its pose have never been surpassed....
To be sure, anyone who has seen [it] has no need
to see anything else by any other sculptor, living or dead.

LIVES OF THE ARTISTS: VOLUME I
GIORGIO VASARI
TRANSLATED BY GEORGE BULL

From: **WORK** *n* (BEFORE 12C)
2 A : ENERGY EXPENDED BY NATURAL PHENOMENA C : THE TRANSFERENCE OF ENERGY THAT IS PRODUCED BY THE MOTION OF THE POINT OF APPLICATION OF A FORCE AND IS MEASURED BY MULTIPLYING THE FORCE AND THE DISPLACEMENT OF ITS POINT OF APPLICATION IN THE LINE OF ACTION

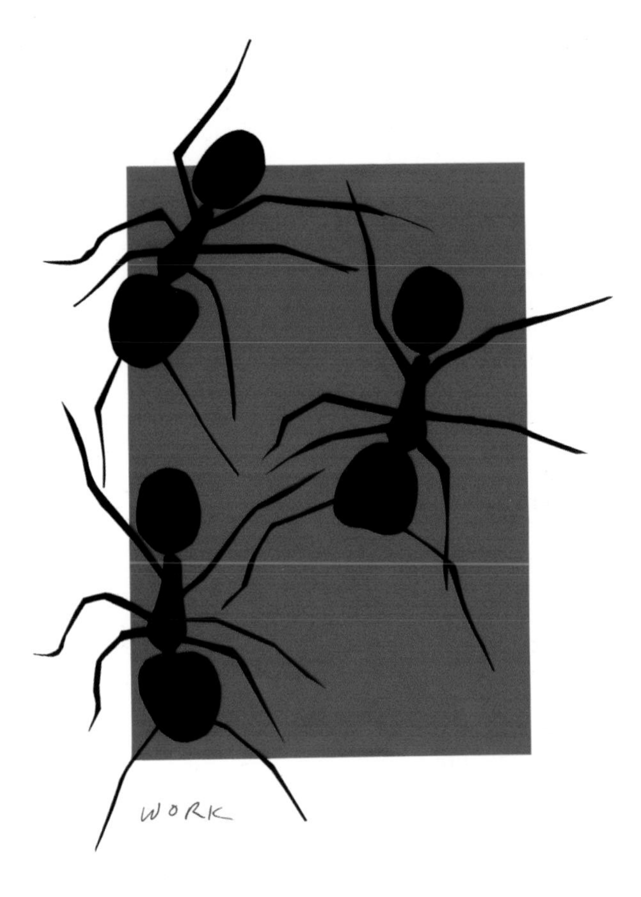
WORK

IF WE PUSH A WALL, we make an effort, but nothing moves—the wall does not gain any kinetic energy. Even though we are applying a force, this force is doing no work. In order for a force to do work on an object, the object must be displaced.

As we push a cart down a straight aisle in a supermarket, for example, the force we apply is doing work. If the force is constant, the work done is simply given by the magnitude of the force times the distance traveled—the *displacement*. To push the cart double the distance would double the work done. To push the cart into the ground and forward at the same time would be described in physics as a *force with a downward component and a forward component*. In this case, only the forward component would be doing work. The downward component is doing no work because the cart has no possibility of vertical motion. Only the component of force along the direction of motion does work. If the cart could move without friction and if the force we apply is constant—and if it is the only horizontal force acting on it—then the cart would accelerate. It would increase its speed at a constant rate. Since the speed increases, its kinetic energy increases. The work done by the force would then be equal to the increase in kinetic energy of the cart. The energy we use to apply the force turns into the kinetic energy of the cart. Whenever a force is doing work, energy is transformed.

In reality, we rarely see carts accelerating down supermarket aisles. Usually carts are moved at a constant speed. If the cart moves at a constant speed, its acceleration is zero, and the total force acting on it must be zero. And so, there must be two opposite horizontal forces acting on the cart: the *force we apply* and the *friction force*. Since the speed of the cart does not increase, its kinetic energy does not increase. The mechanical energy provided by the work of the force we apply is turned into heat by the work of the friction force. Both forces "do" work on the cart: our force does positive work, it provides energy to the cart, while the friction force does negative work, it takes this energy and turns it into heat. The amount of work is equal and opposite because the kinetic energy of the cart doesn't change.

See: Acceleration, Energy, Force, Friction

$$W = F_x \Delta x$$

[Eq. 14] ***The work (W) done by a constant force is equal to the product of the component of the force along the direction of motion (F_x) times the displacement (Δx).***

THINK ABOUT IT: *Conservative Forces*

When a pendulum swings down from a higher position to a lower one, the *gravitational force* does work on it. The amount of work done is proportional to the difference in *gravitational potential energy* that the pendulum loses, which depends exclusively on the height of the object. As a book falls from a table to the floor, the work done by the gravitational force depends on the height of the table alone. When the work done by a force on a body only depends on the initial and final positions of the body, we call the force *conservative*. The gravitational force is a conservative force. The *force of friction* is not conservative: as we slide a block of wood on the floor, the work done by the friction force depends on the distance traveled—that is to say, on the particular trajectory it follows from its initial to its final position.

READ ABOUT IT: *Careers & The Origin of Little Things*

Working [1974] by Studs Terkel — Pulitzer Prize-winner Terkel ("The Good War") offers a rich vision of American life via compelling first-person interviews with people involved in various professions.

Panati's Extraordinary Origins of Everyday Things [1987] by Charles Panati — This delightful reference book is perfect for those who tend to have friendly disagreements about the origins of things. Panati discusses everything from death traditions (e.g., Asia, 50,000 years ago) to Velcro, Vicks VapoRub, pop-up toasters, zippers, spoons, glass windows, sedatives, and even Halloween.

TALK ABOUT IT: *Beyond Physics—Work & Community*

Organizing a work force ... Hierarchy versus teamwork or collaboration ... How does hierarchy (literally, "a division of angels") put a premium on dominance and rank as opposed to cooperation, creativity, and progressiveness? ... How do organizations that assign fixed responsibilities to fixed job descriptions suffer and/or profit from a lack of accountability among employees (e.g., "it's not my job," "it's not my fault")? ... Does such an organization free itself from responsibility in the event that something goes wrong (i.e., if only one link of the chain can be held accountable, then the rest is absolved from blame)? ... Does this management style benefit the consumer? ... How might businesses, institutions, and families avoid needless politicization and strive for mutual loyalty and meaningful interaction? ... Think about school rank (class levels, clubs, teams, student government, etc.) ... What is the satisfaction of such temporary gains when there is always someone more highly situated? ...

***Le blé noir (The Buckwheat Harvest)* [1888]**
by Emile Bernard
Oil on canvas
Private Collection

Bernard was a post-Impressionist painter and theorist whose style was more expressive of feeling than of detail. Like Gauguin (See: *Friction*) and Denis (See: *Uncertainty*), the artist was a member of the Pont-Aven school. The three hunched *workers* in the upper right corner of the painting are stylized to a fantastic extreme—one needs no other information than these bits of visual punctuation to understand action at a depth. Bernard later published his personal correspondence with fellow artists in his magazine, *La Rénovation Esthétique* [1904], helping to establish an appreciation of modern art.

***The Lamp of Beauty* [1843-1860] by John Ruskin**

It is especially fitting to end with this quote, in that the artists and scientists described in this book have enriched the human journey with their *work*. Through efforts both grave and glorious (i.e., their "utmost"), we find ourselves in possession of a legacy of ideas and information that will challenge, comfort, and guide us into the coming centuries. Ruskin's writing is unusually opinionated, but one is made all the better for the introduction to his strong ideas.

***Lives of the Artists: Volume I* [1550] by Giorgio Vasari**
Translated by George Bull

This story of moving the *David* through the streets of Florence provides extraordinary examples of *work* all around, from the artist who made the perfect sculpture, to the brothers who devised the ingenious transport plan, to Vasari, who—thankfully—committed it all to paper.

FOOTNOTES & BIBLIOGRAPHY

Introduction —
1 **Levin**, Janna, *How the Universe Got Its Spots: Diary of a Finite Time in a Finite Space*. — **Baeyer**, Hans Christian von, *Warmth Disperses and Time Passes: The History of Heat*. — **Capra**, *The Turning Point: Science, Society, and the Rising Culture*. — **Gell-Mann**, Murray, *The Quark and the Jaguar: Adventures in the Simple and the Complex*. — **Arnheim**, Rudolf, *Entropy and Art: An Essay on Disorder and Order*. — **Hofstadter**, Douglas, *Metamagical Themas: Questing for the Essence of Mind and Pattern*. — **Dennett**, Daniel, *Darwin's Dangerous Idea: Evolution and the Meanings of Life*. — **Smolin**, Lee, *Three Roads to Quantum Gravity*.
2 **Thernstrom**, Melanie. 2004. The Writing Cure. *The New York Times*, 18 April, final edition.
Jeynes, William. 2004. "Parental Involvement and Secondary School Student Educational Outcomes: A Meta-Analysis." The Harvard Family Research Project *Evaluation Exchange*, Winter.

Acceleration —
1 **McLuhan**, Marshall. *Understanding Media: The Extensions of Man*. Cambridge: The MIT Press, 1994, page 241.
Hagen, Rainer, and Rose-Marie Hagen. *What Great Paintings Say: Old Masters in Detail*. Köln: Benedikt Taschen Verlag, 2000, page 493.
Angle —
2 **Harrison**, Charles and Paul Wood, eds. *Art in Theory: 1900-1990: An Anthology of Changing Ideas*. Oxford: Blackwell Publishers Ltd., 1992, page 37.
Angular Velocity —
3 ***Beowulf***, Seamus Heaney, trans. New York: W. W. Norton & Company, 2000, page xi.
Antimatter —
Selz, Jeans. *Matisse*. New York: Crown Publishers, 1990, page 90.
Chaos —
4 **Janson**, H. W., and Anthony F. Janson. *History of Art: The Western Tradition*. Revised Sixth Edition. Upper Saddle River: Pearson Education, 2004, page 438.
Electricity —
5 **Franklin**, Benjamin. *The Autobiography of Benjamin Franklin*. Leonard W. Labaree, Ralph L. Ketcham, Helen C. Boatfield, and Helene H. Fineman, eds. New Haven: Yale University Press, 1964, page 244.
Walker, John. *The National Gallery of Art, Washington, D.C.* New York: Harry N. Abrams, Inc., 1976, page 575.
Energy —
6 **Everdell**, William R. *The First Moderns*. Chicago: University of Chicago Press, 1997, pages 21-24.
Entropy —
Reti, Ladislao, ed. *The Unknown Leonardo*. New York: Harry N. Abrams, Inc., 1990, page 31.
Equation —
7 **Janson**, pages 828-829.
Field —
Janson, page 852.
Force —
8 **Woolf**, Virginia, Commentary from *Aurora Leigh*. Browning, Elizabeth Barrett. Margaret Reynolds, ed. Ohio: Ohio University Press, 1992, page 443.
Richardson, John. A. *A Life of Picasso: Volume I, 1881-1906*. Toronto: Random House of Canada, 1991, page 227.
Friction —
Silverman, Debora. *Van Gogh and Gauguin: The Search for Sacred Art*. New York: Farrar, Straus and Giroux, 2000, page 375.
Gravity —
9 **d'Arcais**, Francesca Flores. *Giotto*. Raymond Rosenthal, trans. New York: Abbeville Publishing Group, 1995, page 139.
10 ibid, page 141.
11 **Becker**, Ernest. *The Denial of Death*. New York: The Free Press, a Division of Simon & Schuster, 2001, page 284.

Heat —
Walker, page 308.
Light —
"The Official Web Site of Tudely, England." <*www.tudeley.org/chagall.html*>
Mass —
12 **Einstein**, Albert and Leopold Infeld. *The Evolution of Physics*. New York: The Free Press, 1967, page xi.
Baldini, Umberto. "Sculpture." *The Complete Work of Michelangelo*. New York: Reynal and Company, page 144.
Measurement —
Baetjer, Katharine, and J. G. Links. *Canaletto*. New York: Harry N. Abrams, Inc., 1989, page 160.
Momentum —
Burrell, Roy. *Oxford First Ancient History*. Oxford: Oxford University Press, 1991, page 77.
Motion —
Kamien, Roger. *Music: An Appreciation*. New York: McGraw-Hill, 2004, pages 251-254.
Orbit —
Janson, page 748.
Nochlin, Linda. *Realism*. London: Penguin Books, 1991, pages 14-15.
Particle —
13 **Hagen**, page 452.
Position —
14 **Wollstonecraft**, Mary. *A Vindication of the Rights of Woman*. New York: Alfred A. Knopf, 1992, page viii.
Pressure —
15 **Mumford**, Lewis. *The City in History: Its Origins, Its Transformations, and Its Prospects*. San Diego: A Harvest Book, 1989, page 566.
Radiation —
16 **Rosenthal**, Alan, ed. *New Challenges for Documentary*. Berkeley: University of California Press, 1988, page 587.
Grosenick, Uta, ed. *Women Artists in the 20th and 21st Century*. Köln: Benedikt Taschen Verlag GmbH, 2001, page 396.
Reflection —
17 **Kemp**, Martin. *The Science of Art: Optical Themes in Western Art from Brunelleschi to Seurat*. New Haven: Yale University Press, 1990, page 107.
Space-time —
Everdell, page 247.
Richardson, page 68.
Spin —
Silverman, page 418.
Uncertainty —
18 **Hofstadter**, page 464.
Wave —
Janson, page 833.
Wave/Particle Duality —
19 **Hawking**, Stephen. *A Brief History of Time: The Updated and Expanded Tenth Anniversary Edition*. New York: Bantam Books, 1998. Original edition, *A Brief History of Time,* Bantam Books, 1988, page 187.

SUGGESTED READINGS

Abbott, Edwin A. *Flatland: A Romance of Many Dimensions*. New York: Penguin Books, 1998. Original edition, Seeley & Co., 1884.

Alton, John. *Painting with Light*. Berkeley: University of California Press, 1995. Original edition, Macmillan Co., 1949.

Arnheim, Rudolf. *Entropy and Art: An Essay on Disorder and Order*. Berkeley: University of California Press, 1974. Original edition, The Regents of the University of California, 1971.

Asimov, Isaac. *Atom: Journey Across the Subatomic Cosmos*. New York: Plume, 1992. Original edition, Dutton, 1991.

Baeyer, Hans Christian von. *Warmth Disperses and Time Passes: The History of Heat*. New York: Modern Library, 1999. Original edition, Random House, as *Maxwell's Demon,* 1998.

Barabási, Albert-László. *Linked: How Everything Is Connected to Everything Else and What It Means for Business, Science, and Everyday Life*. New York: Plume, 2003.

Barbour, Julian. *The End of Time: The Next Revolution in Physics*. New York: Oxford University Press, 2001. Original edition, Oxford University Press, 1999.

Barron, Frank, Alfonso Montuori, and Anthea Barron, eds. *Creators on Creating: Awakening and Cultivating the Imaginative Mind*. New York: Jeremy P. Tarcher/ Putnam, 1997.

Barthes, Roland. *Mythologies*. Annette Lavers, trans. New York: The Noonday Press, 1990. Original edition, Editions du Seuil, 1957.

Becker, Robert O., and Gary Selden. *The Body Electric*. New York: Morrow, 1985.

Benjamin, Walter. *Illuminations: Essays and Reflections*. Hannah Arendt, ed. Harry Zohn, trans. New York: Schocken Books, 1969. Original edition, Suhrkamp Verlag, 1955.

Berger, John. *Ways of Seeing*. New York: Penguin Books, 1977. Original edition, BBC Television and Penguin Books Ltd., 1972.

Bierce, Ambrose. "An Occurrence at Owl Creek Bridge." *Civil War Stories*. New York: Dover Publications, Inc., 1994. Original edition, *The Collected Works of Ambrose Bierce, Volumes I and II*, The Neale Publishing Company, 1909.

Bourg, David M. *Physics for Game Developers*. Cambridge: O'Reilly & Associates, Inc., 2002.

Briggs, John. *Fractals: The Patterns of Chaos*. New York: Touchstone, 1992.

Brooks, Rodney A. *Flesh and Machines: How Robots Will Change Us*. New York: Vintage Books, 2003. Original edition, Pantheon Books, 2002.

Burke, James. *Connections*. New York: Little, Brown and Company, 1995. Original edition, Macmillan London Limited, 1978.

Caldicott, Helen. *Nuclear Madness: What You Can Do*. New York: W. W. Norton & Company, Inc., 1994. Original edition, W. W. Norton & Company, Inc., 1978.

Capra, Fritjof. *The Turning Point: Science, Society, and the Rising Culture*. New York: Bantam Books, 1988. Original edition, Simon & Schuster, 1982.

——. *The Web of Life: A New Scientific Understanding of Living Systems*. New York: Anchor Books, 1997. Original edition, Anchor Books, 1996.

Carson, Rachel. *Silent Spring*. New York: Mariner Books, 2002. Original edition, Houghton Mifflin Company, 1962.

Chomsky, Noam. *The Common Good*. Tuscon: Odonian Press, 1998.

Clawson, Calvin C. *The Mathematical Traveler: Exploring the Grand History of Numbers*. New York: Perseus Publishing, 2003.

Cohen, I. Bernard, and Richard S. Westfall. *Newton: Texts, Backgrounds, Commentaries*. New York: W. W. Norton & Company, Inc., 1995.

Cole, K. C. *The Hole in the Universe: How Scientists Peered over the Edge of Emptiness and Found Everything*. San Diego: A Harvest Book, 2001.

The Corporation. Directed by Mark Achbar and Jennifer Abbott. Written by Joel Bakan. Vancouver, B.C.: Big Picture Media Corporation, 2004.

Cosmos by Carl Sagan: Collector's Edition DVD Boxed Set. Produced by Adrian Malone. Studio City, C.A.: Cosmos Studios, 2000. Original production, PBS, 1979.

Cousteau, Jacques-Yves, with Frédéric Dumas. *The Silent World (National Geographic Adventure Classics)*. Washington, D.C.: National Geographic Society, 2004. Original edition, Hamish Hamilton, London, 1953.

Dennett, Daniel C. *Darwin's Dangerous Idea: Evolution and the Meanings of Life*. New York: A Touchstone Book, 1995.

Dogtown and Z-Boys. Directed by Stacy Peralta. New York: Sony Pictures, 2001.

Ehrlich, Robert. *Why Toast Lands Jelly-Side Down: Zen and the Art of Physics Demonstrations*. Princeton: Princeton University Press, 1997.

Einstein, Albert. *Ideas and Opinions*. New York: Modern Library, 1994. Original edition, Crown Publishers, 1954.

Everest. Directed by Greg MacGillivray. Laguna Beach, C.A.: MacGillivray Freeman Films, 1998.

Falck-Ytter, Harald. *Aurora: The Northern Lights in Mythology, History and Science*. Hudson: Bell Pond Books, 1999. Original edition, Verlag Freies Geistesleben, 1983.

Ferris, Timothy. *Coming of Age in the Milky Way*. New York: Perennial, 2003. Original edition, William Morrow, 1988.

Feynman, Richard P. *Six Easy Pieces: Essentials of Physics Explained by Its Most Brilliant Teacher*. Cambridge: Perseus Books, 1995. Original edition, *The Feynman Lectures on Physics*, Addison-Wesley Publishing Co., Inc., 1963.

——. *Six Not-So-Easy Pieces: Einstein's Relativity, Symmetry, and Space-Time*. Cambridge: Perseus Books, 1997. Original edition, *The Feynman Lectures on Physics*, Addison-Wesley Publishing Co., Inc., 1963.

——. *QED: The Strange Theory of Light and Matter*. Princeton: Princeton University Press, 1988. Original edition, Princeton University Press, 1985.

Feynman, Richard P, Robert B. Leighton, and Matthew L. Sands. *The Feynman Lectures on Physics: Commemorative Issue*. Reading: Addison-Wesley, 1989. Original edition, 1963.

Fölsing, Albrecht. *Albert Einstein: A Biography*. Ewald Osers, trans. New York: Penguin Books, 1998. Original edition, Suhrkamp Verlag, 1993.

Foucault, Michel. *This is Not a Pipe*. James Harkness, trans. Berkeley: University of California Press, 1983. Original edition, Montpellier, 1973.

Fraser, Gordon. *Antimatter: The Ultimate Mirror*. Cambridge: Cambridge University Press, 2002. Original edition, Cambridge University Press, 2000.

Friedberg, Richard. *An Adventurer's Guide to Number Theory*. New York: Dover Publications, Inc., 1994. Original edition, McGraw-Hill Book Company, 1968.

Gamow, George. *Mr. Tompkins in Paperback*. Cambridge: Canto, 2003. Original edition, Cambridge University Press, 1965.

——. *Thirty Years that Shook Physics: The Story of Quantum Theory*. New York: Dover Publications, Inc., 1985. Original edition, Doubleday, 1966.

Gardner, Martin. *Relativity Simply Explained*. New York: Dover Publications, Inc., 1997. Original edition, Macmillan Company, 1962.

Gelbspan, Ross. *Boiling Point: How Politicians, Big Oil and Coal, Journalists, and Activists Have Fueled the Climate Crisis—and What We Can Do to Avert Disaster*. New York: Basic Books, 2004.

Gell-Mann, Murray. *The Quark and the Jaguar: Adventures in the Simple and the Complex*. New York: Henry Holt and Company, 1994.

Gibilisco, Stan. *Physics Demystified: A Self-Teaching Guide*. New York: McGraw-Hill Companies, 2002.

Ginsburg, Faye D., Lila Abu-Lughod, and Brian Larkin, eds. *Media Worlds: Anthropology on New Terrain*. Berkeley: University of California Press, 2002.

Gleick, James. *Chaos: Making a New Science*. New York: Penguin Books, 1988.

——. *Faster: The Acceleration of Just About Everything*. New York: Vintage Books, 2000. Original edition, Pantheon Books, 1999.

Goethe, Johann Wolfgang von. *Theory of Colours*. Cambridge: MIT Press, 1994. Original edition, John Murray, London, 1840.

Gonick, Larry, and Art Huffman. *The Cartoon Guide to Physics*. New York: Harper-Perennial, 1991.

Greene, Brian. *The Elegant Universe: Superstrings, Hidden Dimensions, and the Quest for the Ultimate Theory*. New York: Vintage Books, 2000. Original edition, W. W. Norton & Company, Inc., 1999.

Gribbin, John. *In Search of Schrödinger's Cat: Quantum Physics and Reality*. New York: Bantam Books, 1984.

Gribbin, John, with Mary Gribbin. *Almost Everyone's Guide to Science: The Universe, Life and Everything*. New Haven: Yale University Press, 2000. Original edition, Weidenfeld & Nicolson, 1998.

Groundhog Day. Directed by Harold Ramis. Culver City, CA: Columbia Pictures, 1993.

Haddad, Leïla, and Alain Cirou. *Mapping the Sky: The Essential Guide to Astronomy*. San Francisco: Chronicle Books, 2003. Original edition, Editions du Seuil, 2001.

Hawking, Stephen. *A Brief History of Time: The Updated and Expanded Tenth Anniversary Edition*. New York: Bantam Books, 1998. Original edition, *A Brief History of*

Time, Bantam Books, 1988.

Heath, T. L. *The Works of Archimedes*. New York: Dover Publications, Inc., 2002. Original edition, Cambridge University Press, 1897.

Hebra, Alex. *Measure for Measure: The Story of Imperial, Metric, and Other Units*. Baltimore: The Johns Hopkins University Press, 2003.

Heifetz, Jeanne. *When Blue Meant Yellow: How Colors Got Their Names*. New York: Henry Holt and Company, Inc., 1994.

Hiroshima-Nagasaki, August 1945. Directed/Produced by Erik Barnouw and Akira Iwasaki. New York: Columbia University Center for Mass Communications, 1970.

Hoffmann, Peter. *Tomorrow's Energy: Hydrogen, Fuel Cells, and the Prospects for a Cleaner Planet*. Cambridge: The MIT Press, 2002.

Hofstadter, Douglas R. *Metamagical Themas: Questing for the Essence of Mind and Pattern*. New York: Basic Books, 1985.

Hogan, Craig. *The Little Book of the Big Bang: A Cosmic Primer*. New York: Copernicus, 1998.

Horowitz, Paul, and Winfield Hill. *The Art of Electronics*. Cambridge: Cambridge University Press, 2001. Original edition, Cambridge University Press, 1980.

Huff, Toby E. *The Rise of Early Modern Science: Islam, China, and the West*. Cambridge: Cambridge University Press, 2003. Original edition, Cambridge University Press, 1993.

Jammer, Max. *Concepts of Mass in Contemporary Physics and Philosophy*. Princeton: Princeton University Press, 2000. Original edition, Harvard University Press, 1961.

Jeans, Sir James. *Physics and Philosophy*. New York: Dover Publications, Inc., 1981. Original edition, Cambridge University Press, 1943.

——. *Science & Music*. New York: Dover Publications, Inc., 1968. Original edition, Cambridge University Press, 1937.

Kaku, Michio. *Hyperspace: A Scientific Odyssey Through Parallel Universes, Time Warps, and the Tenth Dimension*. New York: Anchor Books, 1995. Original edition, Oxford University Press, 1994.

Kasper, Joseph E., and Steven A. Feller. *The Complete Book of Holograms: How They Work and How to Make Them*. New York: Dover Publications, Inc., 2001. Original edition, John Wiley & Sons, Inc., 1987.

Kendall, Richard. *Van Gogh's Van Goghs: Masterpieces from the Van Gogh Museum in Amsterdam*. New York: Harry N. Abrams, Inc., 1998.

Keyes, Daniel. *Flowers for Algernon*. New York: A Harvest Book, 2004. Original edition, 1959.

Kuhn, Thomas S. *The Structure of Scientific Revolutions*. Chicago: The University of Chicago Press, 1996. Original edition, The University of Chicago Press, 1962.

Lao Tzu. *Tao Teh Ching*. John C. H. Wu, trans. Boston: Shambhala Publications, Inc., 1990. Composed approximately 500 B.C.E.

Levin, Janna. *How the Universe Got Its Spots: Diary of a Finite Time in a Finite Space*. New York: Anchor Books, 2003.

Lightman, Alan. *Einstein's Dreams*. New York: Warner Books, Inc., 1994. Original edition, Pantheon Books, 1993.

Lind, David, and Scott P. Sanders. *The Physics of Skiing: Skiing at the Triple Point*. New York: Springer-Verlag, 2004. Original edition, Springer-Verlag, 2003.

Livingstone, Margaret. *Vision and Art: The Biology of Seeing*. New York: Harry N. Abrams, Inc., 2002.

Lucie-Smith, Edward. *The Thames & Hudson Dictionary of Art Terms*. London: Thames & Hudson Ltd., 2003.

Mandelbrot, Benoit B. *The Fractal Geometry of Nature*. New York: W. H. Freeman and Company, 2004. Original edition, 1977.

March, Robert H. *Physics for Poets*. New York: McGraw-Hill Companies, 2002. Original edition, McGraw-Hill, Inc., 1970.

Mishkin, Andrew. *Sojourner: An Insider's View of the Mars Pathfinder Mission*. New York: The Berkley Publishing Group, 2003.

Mumford, Lewis. *The City in History: Its Origins, Its Transformations, and Its Prospects*. San Diego: A Harvest Book, 1989. Original edition, Harcourt Brace & Company, 1961.

Nagyszalanczy, Sandor. *Tools Rare and Ingenious: Celebrating the World's Most Amazing Tools*. Newtown: The Taunton Press, 2004.

Newton, Isaac. *The Principia: Mathematical Principles of Natural Philosophy*. Berkeley: University of California Press, 1999. Original edition, 1687.

No Nukes. Directed by Daniel Goldberg, Anthony Potenza, and Julian Schlossberg.

Warner Brothers, 1980. Filmed, September 19-23, 1979. "The MUSE No Nukes Concerts." New York: Madison Square Garden.

Nova: The Elegant Universe. Directed by Joseph McMaster. Boston: WGBH Educational Foundation, 2003.

Panati, Charles. *Panati's Extraordinary Origins of Everyday Things*. New York: Harper- Perennial, 1989. Original edition, Harper & Row Publishers, 1987.

Peat, F. David. *From Certainty to Uncertainty: The Story of Science and Ideas in the Twentieth Century*. Washington, D.C.: Joseph Henry Press, 2002.

Penrose, Roger. *The Emperor's New Mind: Concerning Computers, Minds and The Laws of Physics*. Oxford: Oxford University Press, 1989.

Petroski, Henry. *The Evolution of Useful Things*. New York: Vintage Books, 1994. Original edition, Alfred A. Knopf, 1992.

——. *Remaking the World: Adventures in Engineering*. New York: Vintage Books, 1999. Original edition, Alfred A. Knopf, 1997.

Pinker, Steven. *How the Mind Works*. New York: W. W. Norton, 1999. Original edition, W. W. Norton, 1997.

Rhodes, Richard. *The Making of the Atomic Bomb*. New York: Touchstone, 1988. Original edition, Simon & Schuster, 1986.

Rosenblum, Robert. *Paintings in the Musée d'Orsay*. New York: Stewart, Tabori & Chang, 1989.

Rosenthal, Alan, ed. *New Challenges for Documentary*. Berkeley: University of California Press, 1988.

Rumsey, David, and Edith M. Punt. *Cartographica Extraordinaire: The Historical Map Transformed*. Redlands: ESRI Press, 2004.

Russell, Bertrand. *ABC of Relativity*. New York: Routledge, 2001. Original edition, George Allen & Unwin, 1925.

Scientific American. *Understanding Cosmology*. Various editors. New York: Warner Books, 2002.

Segrè, Gino. *A Matter of Degrees: What Temperature Reveals About the Past and Future of Our Species, Planet, and Universe*. New York: Penguin, 2003. Original edition, Viking, 2002.

Shute Nevil. *On the Beach*. New York: Ballantine Books, 1974. Original edition, William Morrow and Company, Inc., 1957.

Smolin, Lee. *Three Roads to Quantum Gravity*. New York: Basic Books, 2001. Original edition, Weidenfeld and Nicolson, 2000.

Sobel, Dava, and William J. H. Andrewes. *The Illustrated Longitude: The True Story of a Lone Genius Who Solved the Greatest Scientific Problem of His Time*. New York: Walker & Company, 2003. Original edition, *Longitude*, Walker & Company, 1995.

Stewart, Ian. *Does God Play Dice?: The New Mathematics of Chaos*. Oxford: Blackwell Publishing Ltd., 2004. Original edition, Blackwell Publishing Ltd., 1989.

——. *What Shape Is a Snowflake?: Magical Numbers in Nature*. Lewes: The Ivy Press Limited, 2001. Original edition, Weidenfeld & Nicolson, 2001.

Taylor, Charles, ed. *The Concise Science Encyclopedia*. New York: Larousse Kingfisher Chambers, Inc., 2001.

Terkel, Studs. *Working: People Talk About What They Do All Day and How They Feel About What They Do*. New York: The New Press, 2004. Original edition, Pantheon Books, 1974.

Thomas, Lewis. *The Lives of a Cell: Notes of a Biology Watcher*. New York: Penguin Books, 1995. Original edition, The Viking Press, 1974.

Thomson, Richard, Phillip Dennis Cate, and Mary Weaver. *Toulouse-Lautrec and Monmartre*. Princeton: Princeton University Press, 2005.

Uman, Martin A. *All About Lightning*. New York: Dover Publications, Inc., 1986.

Waldman, Gary. *Introduction to Light: The Physics of Light, Vision, and Color*. New York: Dover Publications Inc., 2002. Original edition, Prentice-Hall, Inc., 1983.

Weinberg, Steven. *Dreams of a Final Theory: The Scientist's Search for the Ultimate Laws of Nature*. New York: Vintage Books, 1994. Original edition, Pantheon Books, 1992.

Wheeler, John Archibald. *Journey into Gravity and Spacetime*. New York: Scientific American Library, 1990.

Zimmerman, Barry E., and David J. Zimmerman. *Why Nothing Can Travel Faster Than Light: And Other Explorations In Nature's Curiosity Shop*. Chicago: Contemporary Books, 1993.

Zinn, Howard. *A People's History of the United States: 1942-Present*. New York: HarperCollins, 2003. Original edition, New York: Longman, 1980.

FEATURED LITERATURE & ART

[All quotes and art reprinted by permission.]

Acceleration — McLuhan, Marshall. *Understanding Media: The Extensions of Man*. Introduction by Lewis H. Lapham. Cambridge: The MIT Press, 1994. Reprinted by permission. Original edition, McGraw-Hill Book Company, 1964. — **Hopper**, Edward. *Nighthawks*, 1942. Oil on canvas, 84.1 x 152.4 cm. Friends of American Art Collection, 1942.51. The Art Institute of Chicago, Chicago. Photograph © 2002, The Art Institute of Chicago, All Rights Reserved. www.artic.edu

Angle — White, Hayden. *Tropics of Discourse: Essays in Cultural Criticism*. Baltimore: Johns Hopkins University Press, 1990. Reprinted by permission. Original edition, Johns Hopkins University Press, 1978. — **Cézanne**, Paul. *Mont Sainte-Victoire*, 1897-1898. Oil on canvas, 81 x 100.5 cm. Collection of Bernhard Koehler, Berlin. The State Hermitage Museum, St. Petersburg. www.hermitagemuseum.org

Angular Velocity — *Beowulf*, Seamus Heaney, trans. Copyright © 2000 by Seamus Heaney. Used by permission of W. W. Norton & Company, Inc. Composed sometime between the middle of the 7th Century and the end of the 10th Century. — **Stieglitz**, Alfred. *Equivalent*, 1929. Gelatin silver print, 12.0 x 9.3 cm. Part purchase and part gift of An American Place, ex-collection Georgia O'Keeffe. George Eastman House, Rochester, N.Y. © 2005 The Georgia O'Keeffe Foundation/Artists Rights Society, (ARS), New York. www.eastmanhouse.org

Antimatter — Wells, H. G. *The War of the Worlds*. New York: Random House, 1988. Reprinted by permission. Original edition, Harper & Brothers, 1898. — **Zukav**, Gary. *The Dancing Wu Li Masters: An Overview of the New Physics*. Copyright © 1979 by Gary Zukav. Reprinted by permission of HarperCollins Publishers Inc. Original edition, William Morrow, 1979. — **Matisse**, Henri. *Venus*, Ailsa Mellon Bruce Fund, Image © 2005 Board of Trustees, National Gallery of Art, Washington, D.C., 1952, paper collage on canvas, 1.012 x .765 (39 7/8 x 30 1/8). www.nga.gov

Chaos — Solzhenitsyn, Aleksandr I. *The Gulag Archipelago 1918-1956: An Experiment in Literary Investigation I-II*. Copyright © 1973 by Aleksandr I. Solzhenitsyn. English Language translation copyright © 1973, 1974 by Harper & Row Publishers, Inc. Reprinted by permission of HaperCollins Publishers Inc. Composed 1958. Original edition, Harper & Row, 1974. — **Uccello**, Paolo. *The Battle of San Romano*, 1450. Tempera on panel, 182 x 320 cm. National Gallery, London/Bridgeman Art Library. www.nationalgallery.org.uk

Color — Gogh, Vincent van. *Dear Theo: The Autobiography of Vincent van Gogh*. Irving Stone, ed. New York: Doubleday, 1958. Reprinted by permission. Original Vincent van Gogh letters from 1873-1890. — **Raphael**. *Madonna del Granduca*, 1505. Oil on panel, 33 x 21 1/2 in (83.5 x 54.7 cm). Palazzo Pitti, Florence/Bridgeman Art Library. www.polomuseale.firenze.it

Electricity — Franklin, Benjamin. *The Autobiography of Benjamin Franklin*. Leonard W. Labaree, Ralph L. Ketcham, Helen C. Boatfield, and Helene H. Fineman, eds. New Haven: Yale University Press, 1964. Reprinted by permission. Original edition, from Benjamin Franklin's original manuscripts, 1771-1788. — **Sloan**, John. *The City From Greenwich Village*, Gift of Helen Farr Sloan, Image © 2005 Board of Trustees, National Gallery of Art, Washington, D.C., 1922, oil on canvas, .660 x .857 (26 x 33 3/4). www.nga.gov

Energy — Dennett, Daniel C. *Consciousness Explained*. Boston: Back Bay Books, 1992. Reprinted by permission. Original edition, Little, Brown & Co., 1991. — **Klimt**, Gustav. *The Kiss*, 1907-1908. Oil on canvas, 70 7/8 x 10 7/8 in (180 x 180 cm). Österreichische Galarie Belvedere, Vienna/Bridgeman Art Library. www.belvedere.at

Entropy — Cervantes, Miguel de. *Don Quixote*. Tobias Smollett, trans. New York: Modern Library, 1976. Reprinted by permission. Original edition, 1605. — **Shelley**, Percy Bysshe. "Ozymandias," *Complete Poems of Percy Bysshe Shelley*. New York: Modern Library Classics, 1994. Reprinted by permission. Original edition, "Ozymandias," appeared in *The Examiner*, no. 524 (Jan. 11, 1818). — **Da Vinci**, Leonardo. *The Last Supper*, 1495-1498. Fresco (post restoration), 15' 2 x 28' 10 in (4.6 x 8.8 m). Santa Maria della Grazie, Milan/Bridgeman Art Library. www.cenacolovinciano.it

Equation — Aristotle. *The Categories: On Interpretations: Prior Analytics, Volume I*. Loeb Classical Library Volume 325, translated by H. P. Cooke, Cambridge, Mass.; Harvard University Press, 1938. Reprinted by permission of the publishers and the Trustees of The Loeb Classical Library. The Loeb Classical Library ® is a registered trademark of the President and Fellows of Harvard College. Composed 350 B.C.E. Original edition, Immanuel Bekker, 1831 — **Mondrian**, Piet. *Composition in Color B, 1917*, 1917. Oil on canvas, 50 x 45 cm (19 3/4 x 17 3/4 in). Kröller-Müller Museum, Otterloo, The Netherlands © 2005 Mondrian/Holtzman Trust c/o hcr@hcrinternational.com

Field — Faulkner, William. *The Sound and the Fury*. New York: Vintage Books, 1987. Reprinted by permission. Original edition, Jonathan Cape & Harrison Smith, Inc., 1929. — **Frankenthaler**, Helen. *Madame Matisse*, 1983. Acrylic on canvas, 60 x 122 1/2 in. Collection of Ellen and Gerald Sigal, Washington, D.C.

Force — Browning, Elizabeth Barrett. *Aurora Leigh*, Margaret Reynolds, ed. Ohio: Ohio University Press, 1992. Reprinted with the permission of Ohio University Press/Swallow Press, Athens, Ohio. Original edition, 1856. — **Picasso**, Pablo © ARS, N.Y. *The Tragedy*, Chester Dale Collection, Image © 2005 Board of Trustees, National Gallery of Art, Washington, D.C., 1903, oil on wood, 1.053 x .690 (41 7/16 x 27 3/16) © 2005 Estate of Pablo Picasso/Artists Rights Society, (ARS), New York. www.nga.gov

Friction — Vasari, Giorgio. *Lives of the Artists: Volume I*, George Bull, trans. Penguin Classics, 1987. Copyright © George Bull, 1965. First published, 1550. — **Gauguin**, Paul. *Vairumati*, 1897. Oil on canvas, 2' 4 3/4 x 3' 1 in. Musée d'Orsay, Paris/Bridgeman Art Library. Giraudon. www.musee-orsay.fr

Gravity — Becker, Ernest. *The Denial of Death*. Copyright © 1973 The Free Press, a Division of Simon & Schuster Adult Publishing Group. Copyright © renewed 2001 Marie H. Becker. All rights reserved. — **Giotto**. *The Deposition* (or *The Lamentation*), 1303-1305. Fresco. Scrovegni (Arena) Chapel, Padua/Bridgeman Art Library. www.cappelladegliscrovegni.it

Heat — Thoreau, Henry David. *Walden and Other Writings*, Brooks Atkinson, eds. New York: Random House, 2000. Reprinted by permission. Original edition, 1854. — **La Tour**, Georges de. *The Repentant Magdalen*, Ailsa Mellon Bruce Fund, Image © 2005 Board of Trustees, National Gallery of Art, Washington, D.C., 1640, oil on canvas, 1.130 x .927 (44 1/2 x 36 1/2); framed: 1.219 x 1.365 (48 x 53 3/4). www.nga.gov

Image — Minh-ha, Trinh T. *Woman, Native, Other*. Indianapolis: Indiana University Press, 1989. Reprinted by permission. — **Wilson**, Edward O., *On Human Nature*. page 92-93, Cambridge, Mass.: Harvard University Press, Copyright © 1978 by the President and Fellows of Harvard College. — ***The Falling Cow Panel***, by Unknown. 13,000 B.C.E. Rock painting of a leaping cow and a frieze of small horses. Caves of Lascaux, Dordogne/ Bridgeman Art Library. www.culture.gouv.fr

Light — *The Letters of Abelard and Heloise*. translated by Betty Radice, Copyright © 1974 by Betty Radice. Used by permission of Penguin Group (USA) Inc. — **Chagall**, Marc. *An Angel*, 1978. Stained glass. All Saints Church, Tudeley, Kent, UK/Bridgeman Art Library. © 2005 Artists Rights Society (ARS), New York /ADAGP, Paris. www.tudeley.org

Mass — Einstein, Albert and Leopold Infeld. *The Evolution of Physics*. New York: The Free Press, 1967. Reprinted by permission from the Albert Einstein Archives, Jewish National University Library. Original edition, Simon & Schuster, 1938. — **Buonarroti**, Michelangelo. *Pietà*, 1499. Marble. St. Peter's, Vatican City/Bridgeman Art Library. mv.vatican.va

Measurement — Eliade, Mircea. "Time and Eternity in Indian Thought." *Man and Time: Papers from the Eranos Yearbooks*. Joseph Campbell, ed. Ralph Manheim and R. F. C. Hull, trans. Princeton: Princeton University Press, 1957. Reprinted by permission. Original edition, *Eranos-Jahrbücher* XVII (1949) and XX (1951), Rhein-Verlag. — **Canaletto**. *The Grand Canal from Carità to the Bacino di S. Marco*, circa 1730. Oil on canvas, 18 7/8 x 31 1/2 in (47.9 x 80 cm). The Royal Collection © 2005, Her Majesty Queen Elizabeth II. www.royal.gov.uk

Momentum — Tocqueville, Alexis de. *Democracy in America*. Edited by J. P. Mayer and Max Lerner. Translated by George Lawrence. English translation copyright © 1965 by Harper & Row Publishers Inc. Reprinted by permission of HarperCollins Publishers Inc. Original edition, Saunters and Otley, 1835. — **Hyde**, Lewis. *The Gift: Imagination and the Erotic Life of Property*. New York: Vintage Books, 1983. Reprinted by permission. Original edition, Random House, 1979. — ***Panel with Striding Lion*** , (One of pair), by Unknown. 604-562 B.C.E. Ceramics. Babylonian. Period of Nebuchadnezzar II. The Metropolitan Museum of Art, Fletcher Fund, 1931. (31.12.2) Photograph © 1995 The Metropolitan Museum of Art, New York, N.Y. www.metmuseum.org

Motion — Rand, Ayn. *Atlas Shrugged*. Copyright © 1957 by Ayn Rand, renewed © 1985. Used by permission of Dutton Signet, a division of Penguin (USA) Inc. Original edition, Random House, 1957. — **Beethoven**, Ludwig van. *Ninth Symphony*, 1824. Original Manuscript, photograph. © 2005 Bettmann/CORBIS.

Orbit — Bettelheim, Bruno. *The Uses of Enchantment: The Meaning and Importance of Fairy Tales*. New York: Vintage Books, 1977. Reprinted by permission. Original edition, Alfred A. Knopf, Inc., 1976. — **Capra**, Fritjof. *The Web of Life: A New*

Scientific Understanding of Living Systems. New York: Anchor Books, 1997. Reprinted by permission. Original edition, Anchor Books, 1996. — **Cassatt**, Mary. *The Child's Bath* (or *The Bath*), 1893. Oil on canvas, 39 1/2 x 26 in. Robert A. Waller Fund, 1910.2. The Art Institute of Chicago, Chicago. Image © The Art Institute of Chicago. www.artic.edu

Particle — Proust, Marcel. "Swann's Way", translated by C. K. Scott Moncrieff, copyright © 1981 by Random House, Inc. and Chatto & Windus, from *Remembrance of Things Past: Volume I* translated by C. K. Scott Moncrieff and Terence Kilmartin. Used by permission of Random House, Inc. Original edition, B. Grasset, 1913. — **Seurat**, Georges. *A Sunday on La Grande Jatte*, 1884-1886. Oil on canvas, 81 3/4 x 121 1/4 in (207.5 x 308.1 cm). Helen Birch Bartlett Memorial Collection, 1926. The Art Institute of Chicago, Chicago/Bridgeman Art Library. www.artic.edu

Position — Buck, Pearl S. *The Good Earth*. Reprinted with permission of Pocket Books, an imprint of Simon & Schuster Adult Publishing Group. Copyright 1931 by Pearl S. Buck. Copyright renewed © 1958 by Pearl S. Buck. Original edition, John Day Company, 1931. — **Wollstonecraft**, Mary. *A Vindication of the Rights of Woman*. New York: Alfred A. Knopf, 1992. Reprinted by permission. Original edition, London: J. Johnson, 1792. — **Giacometti**, Alberto. *Walking Man II*, Gift of Enid A. Haupt, Image © 2005 Board of Trustees, National Gallery of Art, Washington, D.C., 1960, bronze, 1.885 x .279 x 1.107 (74 1/4 x 11 x 43 5/8). www.nga.gov

Power — Berger, John. *Ways of Seeing*. Copyright © 1972 by John Berger. Used by permission of Viking Penguin, a division of Penguin Group (USA) Inc. Original edition, BBC Television and Penguin Books Ltd., 1972. — **Machiavelli**, Niccolò. *The Prince*. A Norton Critical Edition, Second Edition translated by Robert M. Adams. Copyright © 1992, 1977 by W. W. Norton & Company, Inc. Used by permission of W. W. Norton & Company, Inc. Composed 1513, published 1532. — **Weyden**, Rogier van der. *Portrait of a Lady*, circa 1460. Oil on panel, 37 x 27.1 cm. National Gallery of Art, Washington, D.C./Bridgeman Art Library. www .nationalgallery.org.uk

Pressure — Mumford, Lewis. *The City in History: Its Origins, Its Transformations, and Its Prospects*. Copyright © 1961 and renewed 1989. Reprinted by permission of Harcourt, Inc. Original edition, Harcourt, Brace & World, 1961. — **Hokusai**, Katsushika. *The Great Wave at Kanagawa, from a Series of Thirty-six Views of Mount Fuji*, 1830-1832. Polychrome woodblock print; ink and color on paper, 10 1/8 x 14 15/16 in (25.7 x 37.9 cm). Tokyo Fuji Art Museum, Tokyo/Bridgeman Art Library. www.fujibi.or.jp

Radiation — Camus, Albert. *The Stranger*, Matthew Ward, trans. New York: Vintage International, 1988. Reprint by permission. Original French edition *L'Etranger*, Librairie Gallimard, 1942. — **O'Keeffe**, Georgia. *Evening Star No. VI*, 1917. Watercolor on paper. 8 7/8 x 12 in. 1997.18.03. Gift of the Burnett Foundation. Georgia O'Keeffe Museum, Santa Fe/Art Resource, N.Y. © 2005 The Georgia O'Keeffe Foundation/Artists Rights Society (ARS), New York. www.okeeffemuseum.org

Reflection — Jeans, Sir James. *Science & Music*. New York: Dover Publications, Inc., 1968. Reprinted by permission. Original edition, Cambridge University Press, 1937. — **Miller**, Alice. *The Drama of the Gifted Child: The Search for the True Self*. New York: Basic Books, 1981. Reprinted by permission. Original edition, *Das Drama des begabten Kindes*, Suhrkamp Verlag, 1979. — **Velazquez**, Diego. *Las Meninas*, 1656-1657. Oil on canvas, 3.18 x 2.76 cm. Museo del Prado, Madrid/Bridgeman Art Library. www.museoprado.mcu.es

Refraction — Thomas, Dylan. *Under Milk Wood: A Play for Voices*. Copyright © 1952 by Dylan Thomas. Reprinted by permission of New Directions Publishing Corp. First performance, Poetry Center of the Young Men's and Women's Hebrew Association, New York City, 1953. — **Young**, Annie Mae. *Blocks and Stripes*. 1970. Cotton, polyester, synthetic blend, 83 x 80 in. Collection of the Tinwood Alliance, Atlanta, G.A. www.quiltsofgeesbend.com

Relativity — Harrison, Charles and Paul Wood, eds. *Art in Theory: 1900-1990: An Anthology of Changing Ideas*. Oxford: Blackwell Publishers Ltd., 1992. Reprinted by permission. Originally published as 'Picasso Speaks' in *The Arts*, New York, May 1923. — **Magritte**, René © ARS, N.Y. *Two Mysteries "Les Deux mystéres,"* 1966. Oil on canvas, 25 9/16 x 31 1/2 in (65 x 80 cm). Private collection/Bridgeman Art Library. James Goodman Gallery, N.Y. © 2005 C. Herscovici, Brussels/Artists Rights Society, (ARS), New York.

Space-time — Gibran, Kahlil. *The Prophet*. Copyright 1923 by Kahlil Gibran and renewed 1951 by Administrators C.T.A. of Kahlil Gibran Estate and Mary G. Gibran. Used by permission of Alfred A. Knopf, a division of Random House, Inc. Original edition, Alfred A. Knopf, Inc., 1923. — **Picasso**, Pablo © ARS, N.Y. *Les Demoiselles d'Avignon*, 1907. Oil on canvas, 8' x 7' 8 in (243.9 x 233.7 cm). Museum of Modern Art, New York/Bridgeman Art Library. Lauros/Giraudon. © 2005 Estate of Pablo Picasso/Artists Rights Society, (ARS), New York. www.moma.org

Spin — Jobim, Antonio Carlos. *Águas de Março (Waters of March)*. Corcovado Music, 2005. Used by permission. Original recording, *Disco de Bolso - O Tom de Tom Jobim e o tal de João Bosco*, Pocket Record, May 1972. — **Gogh**, Vincent van. *Wheatfield with Crows*, 1890. Oil on canvas, 19 7/8 x 40 9/16 in (50.5 x 103 cm). Vincent Van Gogh Museum, Amsterdam/Bridgeman Art Library. www.vangoghmuseum.com

Torque — Du Bois, W. E. B. *The Souls of Black Folk*. New York: Barnes & Noble Classics, 2003. Reprinted by permission. Original edition, A. C. McClurg & Co., 1903. — **Hemingway**, Ernest. *The Sun Also Rises*. Reprinted by permission. Copyright 1926 by Charles Scribner's Sons. Copyright renewed 1954 by Ernest Hemingway. Original edition, Charles Scribner's Sons, 1926. — **Toulouse-Lautrec**, Henri de. *Jane Avril*, 1899. Lithograph, printed in color, 22 1/16 x 14 3/16 in. Gift of the Baldwin M. Baldwin Foundation. San Diego Museum of Art, San Diego/Bridgeman Art Library. www.sdmart.org

Uncertainty — Shakespeare, William. *The Tragical History of Hamlet Prince of Denmark*. A. R. Braunmuller, ed. New York: Penguin Books, 2001. Reprinted by permission. Original edition, 1601. — **Denis**, Maurice. *The Green Trees or Beech Trees in Kerduel*, 1893. Oil on canvas, 0.46 x 0.43 cm. Musée d'Orsay, Paris/Bridgeman Art Library.

Vector — Homer. "Book 21: Odysseus Strings His Bow", *The Odyssey*, translated by Robert Fagles. Copyright © 1996 by Robert Fagles. Used by permission of Viking Penguin, a division of Penguin Group (USA) Inc. Composed between the late 8th and early 7th Century B.C.E. — **David**, Jacques-Louis. *The Death of Marat*, 1793. Oil on canvas, 65 x 50 1/2 in (165 x 128.3 cm). Musées Royaux des Beaux-Arts de Belgique, Brussels/Bridgeman Art Library. www.fine-arts-museum.be.

Velocity — Brown, Claude. *Manchild in the Promise Land*. Reprinted with permission of Scribner, an imprint Simon & Schuster Adult Publishing Group. Copyright © 1965 by Claude Brown. Original edition, Macmillan, 1965. — **Motherwell**, Robert. *No. 7*, 1970. From the Basque Suite. Screenprint on paper, 717 x 560 mm. Presented by Rose and Chris Prater through the Institute of Contemporary Prints, 1975. Tate Gallery, London/Art Resource, N.Y. © Dedalus Foundation, Inc./VAGA, N.Y. www.tate.org.uk

Wave — Vertov, Dziga. *Kino-Eye: The Writings of Dziga Vertov*. Annette Michelson, ed. Kevin O'Brien, trans. Berkeley: University of California Press, 1984. Reprinted by permission. Original edition, "Very Simple Slogans" from *On the Paths of Art*, Moscow; Proletkul't, 1926. — **Lawrence**, Jacob. *The Migration of the Negro, Panel 3*. 1940-41. Casein tempera on hardboard, 12 x 18 in (30.5 x 45.7 cm). The Phillips Collection, Washington, D.C. © 2005 Gwendolyn Knight Lawrence/Artists Rights Society (ARS), New York. www.phillipscollection.org

Wave/Particle Duality — Zukav, Gary. *The Dancing Wu Li Masters: An Overview of the New Physics*. Copyright © 1979 by Gary Zukav. Reprinted by permission of Harper Collins Publishers Inc. Original edition, William Morrow, 1979. — **Miró**, Joan. *A star caresses the breast of a negress. (peinture-poéme)*. 1939. Oil on canvas. 129.5 x 194.3. Tate Gallery, London/Art Resource, N.Y. © 2005 Successió Miró/Artists Rights Society (ARS), New York/ADAGP, Paris. www.tate.org.uk

Work — Ruskin, John. *The Lamp of Beauty: Writings on Art*. Copyright © Phaidon Press Limited, 1959, 1995, www.phaidon.com. Original edition, 1843-1860. — **Vasari**, Giorgio. *Lives of the Artists: Volume I*, George Bull, trans. Penguin Classics, 1987. Copyright © George Bull, 1965. First published, 1550. — **Bernard**, Emile. *The Buckwheat Harvest "Le blé noir,"* 1888. Oil on canvas, 72 x 92 cm. © Josefowitz Collection, New York/Bridgeman Art Library/Peter Willi.

Dictionary Definitions — Used by permission. From *Merriam-Webster's Collegiate® Dictionary, Eleventh Edition*. © 2004 by Merriam-Webster, Incorporated. www.merriam-webster.com

INDEX

ORDER FORM

Categories—On the Beauty of Physics
ISBN: 0-9740266-3-8
___ $24.95 Hardcover

SEND ORDERS TO:

Name___________________________________
Address_________________________________
City______________State_________Zip________
Country_________________________________
Phone__________________________________
E-mail__________________________________

Vernacular Press
560 Broadway, Suite 509
New York, NY 10012
212-343-9074
212-343-3895 fax
orders@vernacularpress.com

FORTHCOMING TITLES FROM VERNACULAR PRESS

Categories—On the Beauty of Biology
Categories—On the Beauty of Chemistry

ALSO AVAILABLE FROM VERNACULAR PRESS

Anthropology of an American Girl
a novel by H. T. Hamann
Fiction
ISBN: 0-9740266-7-0 $24.95 US Hardcover
ISBN: 0-9740266-5-4 $12.95 US Paperback

L'auteur—Writing Journal
ISBN: 0-9740266-1-1 $10.95 US Paperback

Please visit us online at
www.vernacularpress.com

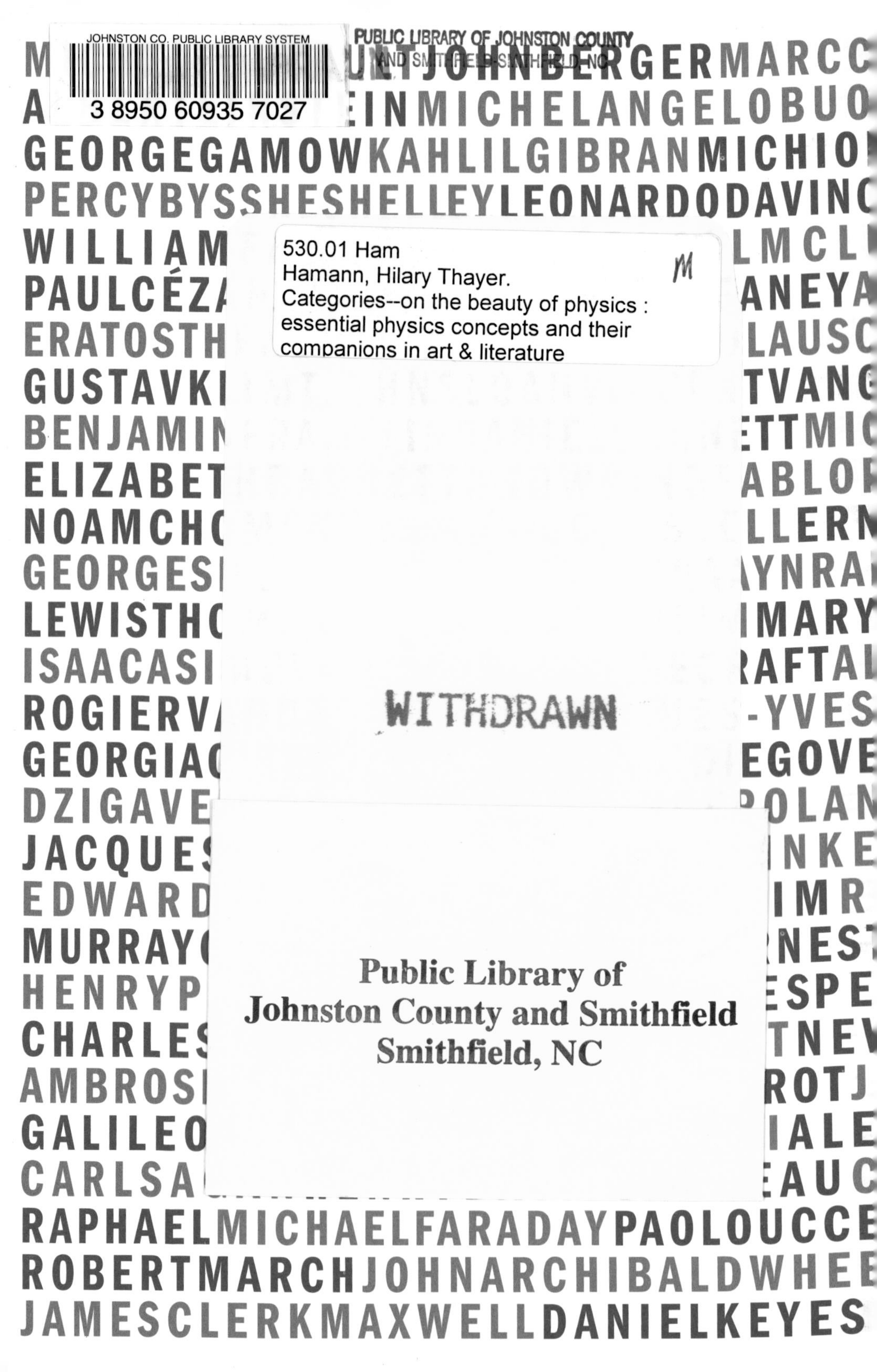